This book commemorates a half century of

industrial design education at Georgia Tech

while celebrating the life and career of

industrial designer and educator Hin Bredendieck

(1904-1995), who served as the program's

director from 1952 to his retirement in 1971.

Georgia Tech is located in Atlanta, Georgia,

and was among the first universities to offer

courses in industrial design beginning in 1940.

BEYOND
BAUHAUS

The Evolving Man-Made Environment

BEYOND
BAUHAUS

The Evolving Man-Made Environment

HIN BREDENDIECK

Unless otherwise noted, images are from the Georgia Tech Archives or from the personal collection of Hin Bredendieck used with permission from his family.

The personal collection eventually will be donated to the existing Bredendieck Collection in the Georgia Tech Archives. Figures 2-4 and 38 were re-photographed by architecture student Erin Howe.

Cover image: stool, wire fabrication project, unidentified student of Hin Bredendieck.

Published by:
Georgia Tech College of Architecture
247 Fourth Street, NW
Atlanta, GA 30332

For more information, contact the College of Architecture Dean's Office at 404.894.3880 or e-mail news@coa.gatech.edu.
Also visit the College of Architecture online at www.coa.gatech.edu.

ISBN 978-0-9823171-1-2
2009939106

Printed in the USA, First Edition

CONTENTS

Hin Bredendieck (1904 - 1995)

foreword

Though Hin Bredendieck had retired several years before I arrived at Tech
in 1978, we talked on occasions and he was still a palpable presence in
the school. I have worked with several major figures whose imagination
was shaped in the *Mitteleuropa* of the 1920s and in my memory of Hin,
he shared with them a prescience and a gravity to his views on the world,
which shaped the character of Industrial Design teaching at Georgia Tech.
We often fail to recognize how much one personality can give substance
to a field of study, and I am most pleased that we are able to celebrate
Professor Bredendieck's exceptional contribution through the generosity
of his student James L. Oliver (BSID 1965, BME 1967).

In a letter to me, Jim Oliver remembers his relationship with his mentor and
teacher and emphasizes the importance of the document that follows not only
to Georgia Tech but to the field of design:

> *Students of Professor Bredendieck who knocked on his office before noon
> were angrily greeted by, "Do not disturb me, I am working on my book."
> In 1989 after I sold my first company and was waiting for my oldest
> daughter to finish high school in Atlanta so we could move to Asheville, NC,
> I contacted the Professor and asked if I could stop by for a visit. I was
> invited back every Tuesday afternoon for the six months before I moved.
> At that time he was in his late eighties but still as sharp as ever. His long
> term memory was very sharp and he vividly described his first classes at
> the Bauhaus and later events when he was an instructor. I thoroughly
> enjoyed this time with him. We also discussed theory of design and both
> of our experiences in industry. After several months I got up the nerve to
> ask about "The Book," which he informed me he was still "refining."
> I asked why he had not published it, and he told me the editors were fools,
> always asking him to change something before they would publish it,
> which he refused to do. I bravely asked if I could have a copy. He printed
> one out from his Mac using the Mac Writer Printer. I promised it was for
> my eyes only and never to be copied or distributed to others. I have read it
> many times since 1989 and have gained from it each time I read it.*

> *Since the 1960s other major universities like Stanford, MIT, and Harvard
> have sponsored research on the theory of design but none of these experts
> have been thinking about design for seventy years like the Professor did.
> I believe his contribution to the study and education of design is a
> very important addition. I am pleased to underwrite the publishing of
> "The Book." I hope all former, current, and future Industrial Design students
> at Georgia Tech will benefit from Professor Bredendieck's seventy years
> of design thought and experience.*

Thank you Jim, and let us reenter the imagination of Hin.

Alan Balfour, Dean
Georgia Tech College of Architecture

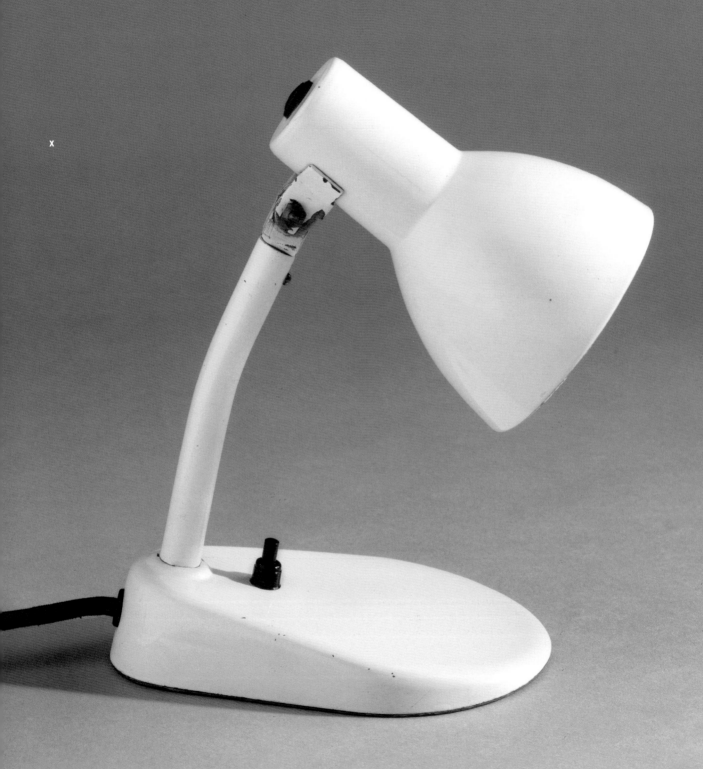

x

editor's preface

Hin Bredendieck and the Roots of the Georgia Tech Industrial Design Program

Beyond Bauhaus: The Evolving Man-Made Environment represents the life's work and passion of Hin Bredendieck, long time director of the Industrial Design Program at Georgia Tech. Born in 1904 in Aurich, Germany, Bredendieck studied at the art academies of Stuttgart and Hamburg before entering the Dessau Bauhaus in 1927 where he was exposed to the Bauhaus philosophy. There, he would begin to make a name for himself as an industrial designer. Working with artist and designer Marianne Brandt (1893-1983) in the metal workshop, he designed many lamps and lighting fixtures including the now ubiquitous Kandem Bedside Table Lamp (fig. i). After graduating in 1930, Bredendieck went to work in the Berlin studios of Bauhaus professors László Moholy-Nagy (1895-1946) and Herbert Bayer (1900-1985). He then worked for B.A.G. Lighting Manufacturing in Turgi, Switzerland, and on lighting for the Corso Theater in Zurich before returning to Germany to work for furniture manufacturer Herman Gautel in Oldenburg. However, the rise of the National Socialist or Nazi Party resulted in the emigration of many artists, including Bredendieck, who left in 1937 at the invitation of László Moholy-Nagy to teach at the New Bauhaus in Chicago, Illinois. Bredendieck led the basic workshop at The New Bauhaus, which later became the Institute of Design and then merged with the Illinois Institute of Design. He would remain in Chicago working as freelance designer and teaching off and on until he accepted the position of director at Georgia Tech in 1952.

Bredendieck's coming to Georgia Tech signified the permanence of the program and Georgia Tech's commitment to Industrial Design. The Department of Architecture, as it was then known, had offered Industrial Design courses since 1940; however, the advent of World War II led to a severe decline in enrollment and the streamlining of course and degree offerings across campus. Bredendieck's former assistant at the New Bauhaus Andi Schiltz had been hired to teach Industrial Design and lead the division in the early 1940s. It was short lived, as ID became a casualty of the war and was no longer an option of study within the architecture curriculum.[i]

After the end of World War II, the push for re-starting the ID Program continued, as was explained in the student magazine *The Georgia Tech Engineer:*

In the field of design a new profession, Industrial Design, is taking an increasingly important place, and, with the advent of industry to the South, the need for the well-trained industrial designer is going to prove of great importance. Before the war such a course was inaugurated at Georgia Tech as Option Three [in Architecture] under a grant from the General Education Board. With the inception of the war this new venture had to be abandoned, and lack of facilities since then has forced the postponing of this option. In spite of a persistent demand for Industrial Design, this division of the department cannot be reinstituted until the plans for a new Architectural Building materialize.[ii]

At the time of the dedication of the East Architecture Building on September 20, 1952, the Industrial Design "lab" and shop were assigned the ground floor of the north wing of the building, and Hin Bredendieck had been hired to lead this new program (fig. ii). The authorization of the Bachelor of Science in Industrial Design (BSID) in the 1950s further symbolized the importance of Industrial Design to Georgia Tech and the region. The first graduates who received the BSID were: James Bayly, John Dunn, David James, Robert Kirby, and William Watkins in 1958 and 1959.

As a teacher, Professor Bredendieck was demanding. He expected the very best out of his students. Many of his former students attribute their success to him—one common theme in talking to alumni of the program is that they did not realize how valuable the education they were receiving was until they left Georgia Tech and began working. One of his early students Irwin Schuster (BS Architecture, ID Option 1957) remembered: "He was a good guy, under-appreciated by me, as a young student… He wore only shades of gray, to make his wardrobe easier to manage." [iii]

Professor Bredendieck brought with him a direct link to the Bauhaus and the modern movement. He studied under the Bauhaus masters, practiced with them, and taught by their side. He represented a shifting ideology that belonged to the modern period; and it would be this *Weltanschauung* or world view that would define his tenure at Georgia Tech. He developed his courses and methods over time or as he put it, "making changes as new knowledge becomes available," and viewed the Bauhaus as one phase of development. The Bauhaus definitely provided the foundation to his methods and ideals. Bauhaus founder Walter Gropius would later describe in a letter to Professor Bredendieck: "You look out for something which I call the science of design, for which the Bauhaus has laid some foundation stones. Your contribution seems to be in the right direction, and I hope you will be able to finish the book you are working on." [iv]

fig. ii East Architecture Building, 1952

This manuscript demonstrates how Professor Bredendieck viewed the progression of his teaching beyond the Bauhaus model; however, the influence of his Bauhaus training is palpable in his writings, his designs, and the program he created. In introducing his system of teaching, he explained, "the approach I suggest is a logical continuation, not so much because I am a product of the Bauhaus but because I see in the early Bauhaus approach the origin of the general ideas expressed here. The Bauhaus accepted the machine as a partner in the task of fulfilling the needs of our society. We must now, as it were, accept the intellect as a worthy partner in design."

Professor Bredendieck retired from Georgia Tech in 1971, having touched the lives of hundreds of graduates, who would go on to become product designers, entrepreneurs, engineers, inventors, artists, and teachers (fig. iii). He remained engaged in the design profession and an outspoken member of the Industrial Design Society of America (IDSA). In a 1987 article in the *IDSA Newsletter,* his influence was recognized: "With Walter Schaer and Eva Pfeil at Auburn, Hin Bredendieck at Georgia Tech, and Walter Baerman at North Carolina State, designers often referred to this educational triangle as the 'New South.' These design educators brought to the South a new design approach which considered user-centered research a prerequisite for intelligent and responsible product development." In 1994, Bredendieck received the IDSA Education Award for his lifelong commitment to design education. He is still recognized by IDSA as one of the discipline's early influences.

In 2009 as the College re-focuses its agenda in industrial design and emphasizes design-based interdisciplinarity at both the graduate and undergraduate levels, it is only fitting that we celebrate a half century of Industrial Design at Georgia Tech through the publication of this manuscript that underscores Professor Bredendieck's commitment to design education. His dedication to excellence in design pedagogy would never lessen throughout his life. Even in the last decade of his life, he stayed on top of what different programs were doing; he continued to participate in design discourse, writing letters to various people and institutions expressing his approval and disapproval of curricular changes; and sought to perfect this manuscript in hopes of its publication. Although I only know Professor Bredendieck through his writings, course materials, designs, family and former students, I cannot help but think he would have celebrated the return of the Common First Year in 1998, the current conversations on design thinking, the College forefronting design within a strong research agenda, and the creation of a Master of Industrial Design degree in 2002.

It has been an honor to have been involved in preparing this manuscript for publication. My overarching goal has been to respect the spirit of the author, whose enthusiasm for his chosen work is captured on each page.

Professor Bredendieck's writing provides us with a better understanding of not just his teaching philosophy but also the roots of the discipline and the Georgia Tech Industrial Design Program. Pedagogical in nature and very much a product of its time, this artifact honors all of our alumni and

fig. iii Bredendieck with ID Student

xiv

especially those who benefitted from the intelligence and engagement of Professor Bredendieck. It is our hope that this project also will mark a critical turning point for the Industrial Design Program—now 166 students strong— as it soars to new heights with fresh perspectives, a vocal and engaged alumni body, and exciting work that reaches into engineering, sciences, and the humanities. His family provided the latest version of the manuscript dated 2002. I made editorial changes for clarity and consistency in the writing and have added more complete references in the form of endnotes. The illustrations were in a large part selected by Bredendieck and mostly have come from his personal collection provided by his son Karl Bredendieck.

His children Dina Zinnes, Gail Fischer, and Karl agreed to have the Georgia Tech College of Architecture publish this manuscript to which they all have contributed in some form. I am thankful for their support and their assistance. Former student of Professor Bredendieck, Jim Oliver (BSID 1965, BME 1967) played a critical role by bringing the manuscript to our attention, introducing me to his children, and, most significantly, by funding the project. Another former student John Hudson (BSID 1964) provided illustrations and encouragement, as he had tried earlier with Claude Hutcheson (BSID 1965) to help move the project along by digitizing many of the images. I also want to thank the College's communications officer Teri Nagel for managing the project and architecture student Erin Howe for assisting with the images and the research.

Leslie N. Sharp, PhD
Assistant Dean, Georgia Tech College of Architecture

· · · ·

notes

i   Frank Graham, "Industrial Design Course Attracts 'Window' Engineers," *The Technique* (21 February 1941): 1; B.B. Holland, "Architecture… Beginning of Tomorrow," *The Georgia Tech Engineer* (May 1950): 10, 58.

ii   Holland, 58.

iii   Irwin Schuster, e-mail to Teri Nagel, COA Communications Officer, 20 July 2008.

iv   Walter Gropius, letter to Hin Bredendieck, undated, Bredendieck Family Collection.

v   "Profile on Walter Schaer," IDSA Newsletter, (Nov/Dec 1987); Walter Schaer was the founder of the IDSA Atlanta Chapter and long-time friend of Hin Bredendieck, Bredendieck Collection, Georgia Tech Library and Archives.

vi   Industrial Design Society of America, "IDSA Design History" section, http://www.idsa.org/whatsnew/sections/dh/edu_awards/1994_Bredendieck.html (accessed 29 June 2009).

BEYOND BAUHAUS

The Evolving Man-Made Environment

HIN BREDENDIECK

Introduction

There are few areas of human activity in our society that have not to some degree felt the impact of a general and unified approach. One area that has held out against this trend is the field of industrial design, in spite of its concern with the vast conglomerate of man-made objects so essential to every person in our society.

Although more and more objects are mass produced, we are only beginning to concern ourselves with the broader implications and impact of designing objects destined for large-scale distribution. Today the interest of the professional designer gravitates around the object to be designed; whatever research is done serves only to solve the problem immediately at hand.

This approach, typical of an evolving profession, has served us up to now. However, it is apparent that, from the viewpoint of further progress, we have reached its limits. A piecemeal concern with numerous individual design problems can no longer contribute to the advancement of the profession. A design approach based largely on the personal make-up of the individual designer is no longer sufficient.

The designer is no longer a mere "problem solver," "aesthetic consultant," or "merchandiser." Neither is he a "craftsman-artist," as proposed by the Bauhaus. The profession of designing is developing into a discipline in its own right, with its own body of knowledge. In many other fields of man's endeavor, at one time in history a comprehensive approach became an essential aspect of the development. Fields such as Geology, Physics, Chemistry, etc. show a definite development over a period of time. The development entailed three aspects: practice, education, and research in a continual feeds back process. This research results in a body of knowledge which feeds back into education and practice. For the industrial designer, however, there is, as of yet, no systematic approach to research. Although computers and vast amounts of pertinent data (concerning human factors, marketing, manufacturing, etc.) are available to today's designers most designers continue to rely largely on their personal make-up, interest, and alertness.[1]

The designer's task, by its very nature, implies change in various aspects of our environment. The designer, involved in such a process, has to cultivate an appropriate dynamic attitude towards the design process itself. The approach of an educator who seeks to evoke the student's creativity must be based on such an attitude, making changes as new knowledge becomes available.

If we think of the next phase in the development of the design profession, we must view the Bauhaus not as a one-time event, but merely as one of the phases in the historical development of dealing with and understanding the man-made environment. The task is to continue the development initiated by the Bauhaus. But to move beyond the Bauhaus of the twenties calls for a scientific organization of knowledge about man-made objects.

In this volume I present such an approach, introducing a holistic outlook in a field that concerned itself previously only with designing isolated objects. This is an initial effort to grasp the complexity of the evolving man-made environment; an attempt to establish a basis for a scientific study of the man-made object, a field of inquiry I call "Objectology." Ultimately we should work towards a comprehensive study of the nature and inter-relationship of objects from spatial and temporal points of view.

Here I have dealt with many aspects only in a cursory fashion. However, in a field where highly personal and rather vague concepts are common, this writing may provide for the searching designer a way out of the present conceptual chaos. It may also serve as an invitation to others to participate in the future development of the field. Perhaps this is what Walter Gropius had in mind when he wrote: "Theory is not a recipe for the manufacturing of works of art, but the most essential element of collective construction; it provides the common basis on which many individuals are able to create together a superior unit of work; theory is not an achievement of individuals but of generations."[2]

In the first chapter, I describe briefly the historical aspects of design education and the role of the Bauhaus. This is followed by a proposal for the classification of the man-made object; the basic attributes of such objects, whose specific characteristics are determined by the designer; the transformation process, that is, how objects come into being; and finally the application of the newly gained knowledge in the education of the designer.

1

A Brief History of Design Education

It is not my intention here to present a detailed account of all aspects of design education. Rather, I propose to trace the overall development, from the pre-Bauhaus period up to the present time, of certain important aspects, specifically, (a) the introductory course, (b) technical training, and (c) the student's conceptual thinking.

Prior to the advent of industrialization in the nineteenth century, the need for consumer goods was met by a variety of craftsmen—the cabinetmaker, potter, blacksmith, and numerous others. These people not only produced but also designed their products. As industrialization accelerated, new factories producing increasingly more goods placed the craftsman, who was not trained to think in terms of machine processes, in a position in which he was unable to compete. No longer having the opportunity to design, he was essentially reduced to the execution of copies of old products.

The origins of design education can be found in the early stages of our industrial era. Initially there were no designers who were able to develop products that could utilize the new means of production. The situation called for a different type of designer. Many of the European art academies came into being to meet the new requirements of rapidly growing industries.

In these schools, a strong emphasis was placed on the visual aspects of design and the ability to sketch and to render. Designs were evolved essentially on the drawing board. An ability to sketch was often a requirement for acceptance into these schools. The strong emphasis on the visual aspects of design may have been a reaction to the technical appearance of industrial products of that time. The all-important element—the student's conceptual thinking—was based on the personal concepts and inclination of the instructor. But the strong emphasis on visualization and the reduction of technical aspects to a secondary role failed to produce the designer required by industry.

This prompted the establishment of a new kind of school in Europe: the arts and crafts school (called the *Kunstgewerbeschule* in Germany). Now not only the visual but the technical aspects of design were to be considered. To meet the need for technical education, the schools instituted various shops equipped with machines. However, these were not an integral part of the curriculum, since students were not required to attend shop courses. The essential point is that in spite of these shops, which indicated a concern for technical training, the educational emphasis remained on the visual aspects of design. No general introductory courses were offered. Development of the students' conceptual thinking continued as before to be grounded in the personal views of their instructors. This was in line with the prevailing notion of the time: creativity is a native ability, and the task of education is to nurture it. Thus the general orientation of the schools, the emphasis on the visual, still resulted in unique, handcrafted products. The lack of technical instruction and the reliance on the personal views of the instructor failed again to educate the kind of designer demanded by the ever-expanding, mass-producing industries.

The Bauhaus Period

It was in 1919, immediately following World War I, that the architect Walter Gropius was appointed to head an existing arts and crafts school in Weimar, Germany. Gropius changed the name of the school to the now famous Bauhaus (Staatliches Bauhaus Weimar). It was his intention to educate designers for industry who would be trained equally to deal with the technical aspects and the formal design of a product—artist-craftsmen in the mold of the masters of various crafts of the pre-industrial era. At the Bauhaus, the work done in other arts and crafts schools of the time was often referred to as "drawing-board designs," the implication being that such work showed little concern for the practical realization of the designs.

Like the arts and crafts schools, the Bauhaus had a number of well-equipped shops. However, the important difference was that these shops were an integral part of the educational program. After attending an introductory course, the *Vorkurs* (basic or preliminary course), the student pursued his educational major by entering one of the shops. The term "shop" is somewhat misleading, since these dealt not only with technical training but, foremost, with design aspects; that is to say, the form of a product. Since there were no instructors at the time who were trained to deal with both technical and formal aspects, each shop had two instructors, a technical master and a form master. The technical master was an accomplished craftsman who had passed both his journeyman and master examinations under the supervision of the Chamber of

Crafts *(Handwerkskammer)*. The form master, a well-known artist who taught design, was in charge of the course. This dual instructorship was considered a temporary arrangement and was to be abandoned once a new generation of artist-craftsmen had been educated.

Quite obviously, the educational emphasis at the Bauhaus was shifted from drawing and rendering to the acquisition of technical knowledge and technical skills. Designs evolved less on the drawing board and more directly in the shops. The designing and the making of a prototype were nearly one and the same process. The strong emphasis on the technical aspects of designing was clearly at the expense of sketching and rendering, which were considered of little importance, perhaps in reaction to the overemphasis on the visual in other schools. No course in sketching and rendering was offered during the early years of the Bauhaus; it was assumed that a photograph of a prototype would be a simpler and better presentation. It should be mentioned in passing, however, that in the later period, in Dessau, courses in freehand drawing were initiated.

In the area of development of a student's conceptual faculty, the educational approach was similar to that of previous schools in that it relied on the personal views of the instructors. However, as a kind of guide, the Bauhaus proclaimed the following tenets:

- The unity of all the arts.

- No distinction between fine and applied art.

- Architecture is the mother of all the arts.

- Get away from the isolated drawing-board design.

- Develop the creative power inherent in everyone.

- Everyone is talented.

- Educate the whole person.

- Accept the machine as being worthy to design for.

To call these tenets essentially rhetorical is not to diminish their importance. Viewed in the

8

context of the time, they were a statement of aims and as such can be considered a first step toward their realization. They reveal a concern with the conceptual aspects of design. Beyond these tenets, the Bauhaus relied on the unstructured views of its instructors. However, a beginning tendency to recognize the broader aspects of design could also be observed.

Some of the student work at the Bauhaus in Weimar differed little from that of other art schools of the time, particularly the ceramic and silver objects.[3] Nevertheless, we see the beginning of an interest in mass production.[4] Important changes occurred in the design of products after the Bauhaus moved from Weimar to Dessau in 1925. These changes constituted a more deliberate trend towards industrial design and away from the previous handicraft approach. It was, I believe, primarily László Moholy-Nagy, head of the metal workshop, and Marcel Breuer, head of the wood workshop, who initiated this trend. Moholy-Nagy established a direct working relationship with certain industries. Many products designed by the students were produced by these industries, in particular lighting fixtures and the tubular furniture designed by Marcel Breuer.[5] Changes were made in the shops and in the courses offered. The Bauhaus in Weimar had seven shops—metal, clay (pottery), glass (stained glass), color (interior and exterior wall painting), textiles (weaving), stone (sculpture), and wood (cabinetmaking); whereas, the Bauhaus in Dessau had six—wood, metal, textiles, paint, and in addition, stage (performance) and print (commercial art). The clay, glass, and stone shops were omitted. The technical training of the student was conducted by a technical master. Since these masters had no design experience, their approach was typical of that used in training craftsmen for the trades.

New conceptual attitudes began to appear in Dessau, which the following two incidents observed by me when I was a student there, may serve to illustrate: A stack of black sheet steel, which had never before been used at the Bauhaus, was ordered by Moholy. When it was delivered, it was placed in the hall next to the door of the metal shop, prompting two passing students of Weimar vintage to remark, "Now it has come to this: using black steel in this school!" The point, of course, is that for some of the older students this was a letdown. They were still thinking in terms of silver or silver-plating, or perhaps brass and copper. Black steel was beneath their artistic consideration. This may have been an isolated remark, but it is in line with the

change in thinking, away from handcrafted products and towards industrial design, that occurred in Dessau.

The second incident reveals the beginning of a trend that has continued to the present day. Moholy had some of his cabinets painted in two-tone gray and gave the following explanation: "The stationary part of the unit is dark gray because it is stationary. The moving parts are light gray because they are movable." It could as well have been the other way around, but this is not the issue here. The important point is not the content of the remark but the tendency to justify the design. Whereas earlier the designer would perhaps simply have said, "That is the way I like it," now he began, and increasingly felt compelled, to justify his design.[6]

The Role of the Faculty at the German Bauhaus

When I attended the Bauhaus in Dessau in 1927-1930, the faculty consisted of the following (named in the order in which they appear in fig.1): Josef Albers, Hinnerk Scheper, Georg Muche, László Moholy-Nagy, Herbert Bayer, Joost Schmidt, Walter Gropius, Marcel Breuer, Wassily Kandinsky, Paul Klee, Lyonel Feininger, Gunta Stölzl, and Oskar Schlemmer.

fig. 2 (opp. page, left)
Hole in Paper Assignment 1
fig. 3 (opp. page, right)
Hole in Paper Assignment 2

Some of these rather impressive and respected personalities already were recognized beyond Europe at the time. The question has been asked, what was the extent of their influence and actual contribution to the education of the students at the Bauhaus and to education in general? Obviously, I can only give a personal and subjective response to this question. In reviewing my years at the Bauhaus and the recent history of design education, I have come to the conclusion that credit should be given primarily to Moholy-Nagy, so far as product design is concerned. It was Moholy who made the move in Dessau away from the handcrafted product towards industrial design and continued this trend in the United States. In giving credit to Moholy, the significant contributions of Marcel Breuer, Herbert Bayer, and Josef Albers, both at the Bauhaus and in their later work, cannot be ignored. However, it was Moholy who (although having fewer actual designs to his credit) assumed the role of a public relations person for the Bauhaus movement. Sometimes resented by close associates as being rather self-serving, he nevertheless made the Bauhaus and its aims better known.

It is not my intention to refer here to every member of the Bauhaus faculty, but some remarks about the participation of the painters Kandinsky and Klee are in order. (The other well-known painter Feininger did not hold a teaching position.) What was their contribution to the education of designers? While not ignoring the importance of these famous personalities, I have somewhat negative feelings about their contribution to the Bauhaus. They afforded prestige to the institution, but it is difficult in retrospect to define a more specific contribution. The material offered in their classes was abstract, and they made no attempt to relate it to the designing of objects. The Bauhaus tenet that proclaims "no distinction between fine and applied art" was discussed among the students on numerous occasions. The question was often asked why Kandinsky and Klee never came into the shops. The students would have liked to receive comments regarding their work in the shops from these respected painters. But shop visits never took place while I was at the school. I can only conclude that the influence of these teachers and their contribution to designing was minimal. This is my feeling in the last analysis, despite Moholy's remark to me that "it was a stroke of genius by Gropius to invite the painters Klee, Kandinsky, and Feininger."

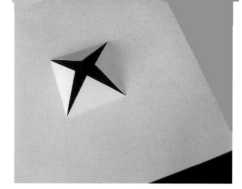

The Introductory Course at the Bauhaus

One of the important innovations in the education of the designer instituted at the Bauhaus was the introductory course, or *Vorkurs*, which has been adopted by many schools here and abroad and has come to be known in the United States as the preliminary course, foundation course, or basic workshop. My personal recollections of this course are vivid.

When I arrived at the Bauhaus in Dessau in the spring of 1927, together with a relatively large group of other freshmen, the introductory course was under the leadership of Josef Albers. I remember my excitement and feeling that I was in the right place, so far as my educational interests were concerned. The atmosphere was different from that of other art schools I had attended. An introductory lecture by Walter Gropius only heightened my expectations, and I anxiously awaited the beginning of my first class with Albers.

The students were assembled as Albers came into the room and introduced himself. He recognized some of them and addressed them by name (possibly recalling them from photos on their application forms). This little opening seemed to relieve some of the tension. Albers lectured on the subject of creativity in general terms for about twenty minutes. Then he proposed an assignment: "Make a hole in paper."

I think we were all without exception visibly baffled, looking from one to the other, wondering what to say or even whether Albers was serious when he proposed such an assignment. Finally someone dared to remark: "One could just bite a hole in a piece of paper." Most likely the student who made this remark thought, as did most of us, that it was a rather silly suggestion. We waited for Albers' response. It was totally unexpected: "See, you have already got one hole," and with this he left the room.

Somehow we began making attempts to produce holes in paper. Someone folded a strip of paper to make it stand up, placing a hole in the upper portion and justifying this hole by saying that it would permit the upper portion of the paper to lean farther over (fig. 2). Another hole was produced by making a crosscut and bending the four triangles outwards, resulting in no loss of paper and the simultaneous gain of a new feature (fig. 3). Still another hole was created by

12

assembling a number of pieces of paper and leaving a hole at the center (fig. 4). It might be obvious, although it was not so to us at the time, that the assignment was not meant to result in definite solutions. Actually, it was a means used by Albers to stimulate the students' creative thinking and to get them away from reacting in a common, routine manner. As I recall, it was the only assignment (if it can be called such) that was given to all students in general.

It was essentially intended to encourage the student to start on his own, to choose his own material and the manner and means of transforming it. Quite early in the course different students were working in a great variety of different materials according to their particular inclinations and the availability of materials and processes. Not only paper but various other materials, such as wood, wire, glass, sheet metal, and plastic, were explored. The results varied widely from student to student and included visual arrangements of different colored glass plates and glass beads, and both simple and intricate structures.[7]

Classes were held every morning, Monday through Friday. As I have intimated, the course was unstructured, with no specific assignments that had to be completed within a certain time, nor were there formal lectures. From time to time, Albers assembled the students to discuss the work in progress. These were rather important gatherings and contributed to the success of the course by providing guidance and further encouragement for the student. During these sessions, Albers explained the ideas that underlay the suggested exercises by formulating catch phrases, such as:

- Economy of materials and tools

- Achieving more with less

- Not copying, but originating

- Exploring the potentials of the materials

- Establishing visual and structural order

Besides such group discussions, Albers talked frequently with each student about his work, making suggestions and helping him to carry a project to a successful conclusion. Each student worked and progressed according to his own pace and inclination, starting and finishing any number of projects. At the end of the semester, the students displayed their work to the entire school and for evaluation by the faculty (fig. 5). The students did not receive grades. On the basis of the work displayed, they were told whether they could continue studying at the Bauhaus.

The Bauhaus approach to the education of designers was introduced in the United States in

fig. 6 (opp. page)
School of Architecture Faculty,
Georgia Tech, 1958

1937 by Moholy-Nagy at the New Bauhaus in Chicago. Moholy, who was my teacher at the Bauhaus in Germany, invited me to participate. The program began with two successive introductory courses. Moholy's first assignment, the making of a so-called tactile chart, was one he had initially introduced at the Bauhaus in Germany.[8] The tactile chart consisted of an arrangement of various textured materials according to the differences in their smoothness or roughness, which were to be experienced by gliding the fingertips over the different surfaces. In the second semester's course, conducted by me, the students began with the so-called paper-cut assignment, originally introduced by Albers at the German Bauhaus.[9] I introduced a change in approach, however. Whereas Albers had given no specific instructions, I stipulated that the problem would be dealt with in three steps. The first was similar to the procedure followed by Albers at the Bauhaus in Germany. The students experimented freely, with no restraints, perhaps even in a playful manner, folding and cutting the paper to produce certain form-structures. The results were then studied in order to understand the underlying principle of each structure. This constituted the second step. The third was the calculated application of some of the discovered principles in the production of a new, well-organized form-structure. Thus the assignment entailed three phases: experimentation, analysis, and exploration. This approach was similarly applied in other assignments devised by me for the introductory course. One of these, the making of a "hand sculpture," was intended to supplement the current over emphasis on the visual aspects of designing, by drawing attention to the importance of man's physical relationship to his environment. Another, the so-called wood-cut assignment, enabled the student to think in terms of machine processes and encouraged him to exploit possible, unknown potentials of materials and processes.[10] Wood was used in this assignment only because it and the necessary machines were available; any other material and process could have been used.

Moholy's Proposed Program for the Post-Introductory Design Courses

Around 1937, Moholy prepared the educational program for the design courses which followed the introductory course. His program was similar to that of the old Bauhaus in that it provided for a number of shops, each focused on one of the materials typically used by the respective trades. This concept had already been opposed by the student body at the old Bauhaus in Germany.

As a member of that body, I was actively involved in seeking the elimination of such a division. In Chicago I again opposed this part of the proposed program, since I believed that the student should have access to all materials and processes; the educational emphasis should be on design, thus training the student to design in any material.

The New Bauhaus was in existence for only two semesters, after which financial difficulties forced it to close. A short time later, Moholy started another school in Chicago, the School of Design, later known as the Institute of Design. It is interesting to note that in this school there was only one shop, in which all materials were dealt with. It was not intended to train the student in technique but essentially served only for the making of try-outs, mock-ups, and prototypes. No specific courses or instruction were offered to familiarize the student with the different materials and processes. In time, however, the students' lack of technical training became apparent; they required too much assistance from their instructors. To alleviate the situation, one session of the regular design course was devoted to acquainting the students with materials and processes. This later became an official, required series of courses, the Material and Technic Course Series, resulting in a more systematic approach.

Further changes in the education of industrial designers were proposed by me at the Georgia Institute of Technology in Atlanta, where I was invited, in 1952, to develop a new Industrial Design Program, one of four options within the School of Architecture (fig. 6). The first year (for all students entering the School of Architecture) consisted of general orientation courses. The industrial design course series began in the second year. Two assignments to which I have already referred, the building of a "hand-sculpture" and the making of a tactile chart, which were introduced at the New Bauhaus in 1937, will serve to illustrate the new changes. In originating the so-called hand-sculpture assignment, my intention had been to draw attention to the importance of man's physical relationship to his environment. I now came to the conclusion that the broader implications of this assignment might not be evident to the student. Therefore, instead of merely assigning the making of a hand-sculpture, I gave a number of lectures that provided the student with tangible information regarding the relationship of man to object, which would be applicable in future design projects. The accompanying assignment now merely served to underline what had been covered in the lectures. Similar changes were made in the tactile-chart assignment. Instead of merely assigning the execution of a chart, a number of lectures informed the students of the various aspects of surfaces in general. The accompanying assignments covered the study of individual aspects of these surfaces. Further information regarding these and other assignments are dealt with in Chapter 7.

At Georgia Tech, the program for industrial designers consisted of two series of courses conducted in parallel: the Design Course Series and the Material and Technic Course Series. It should be emphasized at the outset that both of these series involved design courses. The latter constituted a further development of the earlier Material and Technic Course Series, hence its designation; from an administrative point of view it was simpler to change the content of the course than its name. The difference between these two series of design courses was in the approach towards the designing of an object. In the Design Course Series, the approach began with the definition of a given situation and proceeded, via analysis and synthesis, towards the development of the means by which the problem could be dealt with. In the Material and Technic Course Series, the approach proceeded in the reverse order. It began with the definition of a given means—i.e., the material and/or process—and then attempted to define appropriate situations in which these would be useful. The former was referred to as the "situation approach" and the latter as the "application approach." More will be said about these approaches in Chapter 5.

A Critical Review of the Introductory Course

More than half a century has passed since the Bauhaus originated the introductory course, often referred to as the Bauhaus approach. Since then, it has become part of the educational program in most schools of design. It can now be asked whether it was, and still is, successful and whether it has had a lasting value for the aspiring design student.

The term "Bauhaus approach," when it refers merely to the exercises of the preparatory course, represents a rather limited interpretation of what actually constituted the Bauhaus approach. Disregarding for the moment the limits of such an interpretation, what is the value and role of the introductory course in today's design education? Historically, the introductory course, or basic workshop, as it is also often called, constituted the first such course in which the primary aim was to develop and facilitate the creative faculty of the student. It was instituted by Johannes Itten in 1918 in Vienna. Gropius called Itten to the Bauhaus in Weimar in 1919, and there the approach became the "backbone of the Bauhaus system."[11] Under Itten, the course consisted primarily of instruction in "composition," employing various materials in two and three dimensions, abstract drawing, and analysis of the art of the old masters.[12] Itten had left the Bauhaus by the time it moved to Dessau, and there it was primarily Josef Albers who developed and enriched the introductory course into a significant and imaginative program. In the second semester, Moholy-Nagy, following a similar approach, introduced such assignments as the "tactile chart" and "space modulator," which became part of the introductory course (basic workshop) offered at the New Bauhaus in Chicago. As I have already said, his approach employed no specific methods or procedures; the student was not required or encouraged to produce "premature practical results."[13] Instead, he was offered an opportunity to experiment freely with a variety of materials and tools. There was a strong emphasis on initiative within a "do-it-yourself" situation; both conventional and unconventional means were employed, often resulting in strikingly new and strange configurations.

The undirected manipulation of materials was expected to result in the student's ability to produce structures, forms, patterns, and relationships that would give him a "sense of accomplishment," and this, in turn, would "show him the power that rests within himself."[14] Theoretically, the student was given the means to realize his own creative potential and simultaneously to develop a criterion of form and structure, an ability to recognize the "worthwhile" within the conglomerate of miscellaneous forms produced. It was expected that he would develop an attitude of flexible ingenuity toward unfamiliar forms, which would enable him to exploit these further. Finally, the approach implied that the attitudes, knowledge, and skill thus acquired would be carried over into purposeful design activity.

The apparent success and broad adoption of the basic workshop program would seem to indicate that it achieves these goals, but there is ample cause for skepticism. It is regrettable that in all these years no adequate studies have been made which gage the actual benefits of the approach for the student. Its values have perhaps been considered too obvious to bear further investigation, or perhaps its nature was so elusive that it escaped evaluation. Certain it is that the overall educational program, of which the basic workshop approach is a part, and only a part, shrouds the extent to which it actually affects the later working life of a designer. Even with regard to the immediate returns within the school situation, the effectiveness of the approach is questionable.

What Value Does the Approach Have?

For lack of a systematic study, we must resort to analysis and personal observation. In the Bauhaus exercises, as in any design process, we are dealing with three basic aspects. First, we have a material, which is to be transformed. Second, we have the forming process, which has the potential of transforming the material. Third, we have the design, which determines the forming process that converts the material into a meaningful object of specific character. These three aspects constitute the transformation process, about which more will be said in Chapter 4.

In every process of this kind, all three aspects are always present. But as designers, our chief concern is with the design, since it determines the outcome of the process. This is the aspect that demands foresight and control. Now, when we consider the nature of the basic workshop exercises, we observe at once that they place dominant emphasis on the material and forming aspects of the transformation process and that the role of design is merely incidental. However, the material and forming aspects can be of interest to us as designers only insofar as they impinge on design.

Whenever we manipulate materials in the process of transforming them, whether or not we are paying attention to a design, some agency is always operating to determine the outcome. But such an agency does not necessarily represent the designer as the manipulator of the materials. He may well be merely the extension of his tools or even an unwitting victim of the numerous accidental events in the process. It is precisely the aim of design education to impart to the student the means for achieving authority and command in order to gain ascendancy over the accidental.

The objection here is not that the basic workshop exercises employ tools and materials. Rather, the concern is that the design aspect is given a secondary place, whereas it should permeate every phase of the procedure. Regardless of what specific procedure is followed, to "design" means to control the factors involved and to deliberately develop form. Such design control can only be obtained by intellectual effort. Although such effort may be facilitated through external means, for example through drawing or the manipulation of materials, designing essentially remains a mental act. The extent to which a student succeeds in his design, therefore, depends largely on his attainment of knowledge and understanding.

These goals cannot be achieved through exercises in which the design aspect plays a merely incidental role. As long as the emphasis is placed predominantly on a free manipulation of materials, the most that can be expected is a vague "feeling" of creativity, largely on the subconscious level. It becomes apparent that this approach, due to its very nature, cannot advance a student's understanding of the design process. But then, too, such a goal is not within the scope of these experiences. As important as it is for the student to gain some measure of control over the design process, there are other goals that design education seeks to achieve. These are initiative, resourcefulness, and willingness to assert oneself. In fact, these are precisely the goals that are often listed as the aim of the exercises. Perhaps here lies the actual value of the basic workshop approach.

The Bauhaus approach assumes that "everyone is talented."[15] But it also realizes that such native ability may be inhibited: "The greatest hindrance to creative work is fear."[16] The "chief function" of the introductory course is thus "to liberate the individual by breaking down conventional patterns of thought."[17] The exercises should help the student to "overcome self-conscious fear."[18] Fear and conventional thought patterns are considered to be the "villains" that impede the students' creative efforts. To overcome these impediments, the student is encouraged to play freely with materials. "Free play in the beginning develops courage."[19]

Fear is well known to creative workers and often constitutes a serious obstacle. Many creatve persons have had to overcome such handicaps in the initial stages of their personal development. From an educational point of view, the manner in which man overcomes fear when left to his own devices is interesting. Initially, perhaps, an attempt is made to ignore such fear, but what ultimately conquers it is long-term experience in numerous situations. Fear is not an inherent characteristic but is due merely to unfamiliarity with a particular situation. The student's fear is likewise due to such unfamiliarity and to the fact that he lacks the experience of being reassured by a positive result. Mere freedom and unrestricted manipulation of materials cannot allay this fear. The results of the exercises may, for the moment, give the student a sense of accomplishment, but this will be of little use in later design work.

If fear constitutes a hindrance to creative work, the only means to alleviate it is to gain experience in a variety of areas. Practical work in the field can, in time, provide such experiences and simultaneously allay possible fears. But education can achieve the same end in a deliberate manner, not through play but through a program that exposes the student to a concentration of practical experiences. Thus the very same means that can give the student some measure of control over the design process also can provide him with the means to overcome fear and lack of resourcefulness. Furthermore, since such an educational program is based on a body of knowledge, it provides a new external frame of reference that can be effective in "breaking down conventional patterns of thought."[20]

The apparent air of "freedom" in which the basic workshop approach lets the student operate is a mere phantom, for he is caught in his own net of ignorance and lack of skill. Freedom exists only where there is choice and control. Of course, play can have a place in the creative process, although not necessarily as a means for reducing anxiety. If it is employed at all, it should perhaps constitute a "random" method: a free manipulation of materials and techniques with little intellectual interference. But in order to be of value, educationally or otherwise, play can only be considered as the initial phase of the design process. It must be followed by an analysis of the results of such play and by a conscious attempt to apply the data thus obtained (not necessarily to "useful" objects alone). Only as one of three phases—play, analysis, and application—can play be justified on an educational and professional level.

Some Personal Observations

When I was a student at the German Bauhaus and took the introductory course, first under Albers and later under Moholy, I enjoyed it very much. In later years, many of my own students made similar comments, although when questioned, they, like me, could not point out any specific gains. Perhaps the often "fascinating" objects developed in the course were taken too much at face value with little regard for the effect the process might have on the students' later work; because "interesting" formal structures and relationships resulted from it, the course's educational value was taken for granted.

As an instructor in design courses that followed the basic workshop approach, I often observed a notable inconsistency in the students' performance compared to their later work. One could not predict whether an excellent student in the basic workshop course, or for that matter a mediocre one, would perform equally well, or badly, in following courses. Other instructors must have made similar observations, for I recall a meeting in which precisely this issue was discussed. The following statement was circulated among the faculty prior to the meeting: "Although the purpose of the introductory course is to allow the student to develop his creative abilities freely and without restriction, in the following semesters, when the focus shifts to practical problems, the slightest restriction becomes an obstacle and the student's creativeness shows a tendency to 'freeze.' In most cases he completely ignores all previous training and falls back on conventional patterns. In reviewing work of students in the third and fourth years, there is a definite loss of sensitivity, even as early as the fourth semester."[21]

Of course, there were students who successfully completed the basic workshop and did equally well in later design work. We are easily tempted to list this on the credit side. However, in an unprejudiced evaluation, we also have to consider the possibility that such success might occur in spite of the basic workshop course.

One reason for the often demonstrated inconsistencies in the students' performances at the two levels of their development is the random nature of the approach itself. Another is clearly the lack of coordination between the basic workshop course and the following design courses. Although the basic workshop was intended to be a preparation for the following design courses, it did not evolve from them. From the very beginning, the course was autonomous to a considerable degree. Initially its autonomous character was less apparent, because the underlying concept that prompted the approach was still somewhat similar to the prevailing design concept of the time. But already at the German Bauhaus, the lack of coordination between the courses was frequently discussed among students.

All of this seems to suggest that the approach was essentially ineffectual, even at the early Bauhaus. But such a criticism must be viewed in the context of the time. It should be emphasized that the introduction of this approach constituted a great advance in design education. Its historic importance cannot be minimized. But we must also consider that, at the time of the early Bauhaus, the general tendency was to educate the craftsman-artist. Today, design education strives to develop an individual who is oriented in an essentially different way. This is imperative, because contemporary professional designing, since it is in closer contact with the dynamic events of industrial development, has changed drastically. So far as the actual design process is concerned, design education has generally responded to these changes, but strangely enough the basic workshop course has remained largely in its original form.

How could an approach with such an auspicious beginning, an approach which in its very essence aimed at facilitating creative processes, fail to keep pace with contemporary development? A number of factors have contributed to the now stagnant character of the approach. The most striking, I believe, is that the approach itself was not guided by the spirit that it sought to instill in the student. While it strove to develop creativity, initiative, and resourcefulness, encouraging the student to overcome prejudice and to relentlessly explore new vistas, it failed to apply these very concepts to the approach itself. Its initially creative and certainly dynamic nature thus regressed to a heedlessly repetitive procedure, increasingly losing its contact with reality.

What Remains?

Of course, the entire residue of the Bauhaus approach introducing the student to design cannot be dismissed as meaningless. In the early phases of the movement, Walter Gropius made a number of statements that today can be considered as the harbingers of a development yet to come: "A corresponding knowledge of theory—which existed in a more vigorous era—must again be established as a basis for practice in the visual arts." This statement is as valid today as it was then. "Theory is not the achievement of individuals but of generations." These and other statements were apparently not made casually; they constituted a deliberate call to action. The task is now the responsibility of those who are concerned with the advancement of contemporary design education. If we do not rise to it, the complaint that Gropius once directed toward the academies will apply equally to us. Self-criticism still needs to be practiced, but the Bauhaus movement itself has already been implicated: "The academies, whose task it might have been to cultivate and develop such a theory, completely failed to do so, having lost contact with reality." [22]

In quoting from the early Bauhaus publications, it should be taken into consideration that some terms may have a different connotation for us today. The term "theory," for instance, seems to have had a vaguer meaning at the time. A mere verbalization was apparently considered a theory. Nowhere is the term (as applied to design education) elucidated. In the Bauhaus catalogue it is implied that such a theory actually existed. But it is my impression that the use of the term merely expressed an intention to develop such a theory. In any case, the Bauhaus unquestionably intended to develop and make use of theories, and this is precisely where the movement foreshadowed our present development. It is the only direction that can free us from the present inert state of affairs.

The Development of New Design Concepts

Design by its very nature is a dynamic process. Whether we consider a specific procedure itself or the effect it has on our environment, we witness a constant change and evolution. Design education, in preparing a student to participate in and promote such a development, must necessarily adopt a correspondingly dynamic approach. This means not only that a creative attitude must be instilled in the student himself, but also that a creative attitude must be maintained toward the very approach employed in his education. Design education should prepare the student realistically and effectively for the practice of designing, but an educational design approach patterned solely on practice in the field no longer constitutes an effective educational method. The dominant emphasis in contemporary design education on problem solving and learning by doing, on the manipulative aspects of designing—that is, on training rather than on knowledge—is no longer acceptable.

Design education, essentially a concentration of practical experiences, makes possible an accelerated accumulation of necessary knowledge. This must increasingly be the goal as we advance in our field. But in order to accomplish such a task, design education has to be different in structure and procedure than the practice of designing. It must be different not only because in practice our main concern is the end-result, whereas in education it is the procedure, but also because we must aim at a deliberate concern with the intellectual aspects of designing. We must not only train the student, we must also educate him.

How Can This Be Accomplished?

In the process of designing and in design education in general, we will become increasingly aware of specific and detailed aspects of procedure and approach. The frequent recurrence of an activity in itself makes one aware (at first tacitly, without any directed effort) of specific aspects of a task. Then one becomes increasingly conscious of details, differences, similarities, and order within the process. Eventually, this initial awareness becomes a form of retainable and transferable knowledge. At some point, one consciously decides to investigate the process systematically. From here, it is a short step to organizing the discovered data into a comprehensive structure and then to developing further means for facilitating the process. It thereby becomes a self-perpetuating system. The process of designing, particularly the way it is taught and practiced in institutions of higher learning, can thus no longer simply be a matter of efficiently producing "well-designed" objects. In order to advance both design and the development of sound professional designers, an approach that is merely "hit-and-run" is no longer tenable.

All of this indicates a strong current in the direction of intellectualization; a move apparently away from art and toward science. About this, many will voice concern. Is it a move away from intuition, feeling, and personal relationship to our work? It is certain that some change has already occurred; but such change does not and never will include the abandonment of intuition and "feeling," since this is impossible. Intuition, however, as indispensable as it is for any advancement, cannot and never has done the job alone.

The intellect has always played an integral part in the development of the new. In the past, in design as well as in art, the intellect played its role rather haphazardly. Now the intellect must be employed consciously, as it has been in science for a long time and with remarkable results. True, much of the effort of science has been directed toward the achievement of physical comfort and the development of destructive (once defensive) tools of war, and little has been directed toward cultural pursuits. But this is precisely the point—too little of man's intellectual power is today used for cultural development. The slow pace in art and design, as compared for instance with technology, is not due to the artists' or designers' failure to use intellect (for they have no choice in the matter), but to their lack of advancement to the stage where they are willing to do so consciously. Whereas at one time in history the artist was in the vanguard of development, representing the intellectual elite of his time, he now occupies a rear position. It is a great fallacy to assume that the scientist works only "scientifically" and the artist only "artistically." The basic difference between science and art is not one of approach but rather one of application. Whatever field man ventures into, he must make use of all his faculties in order to advance. The designer and artist, like the scientist, can no longer rely merely on inherent intellect. They have to make a concerted effort to develop the means that will guarantee the full use and further development of both intellect and intuition. There can be no question that the intellect will fructify man's intuitiveness and his intuitiveness will fructify his intellect.

Perhaps the time has come to move closer to the early proclamation of the Bauhaus, "no distinction between fine and applied art," to Gropius' call for a "grammar of design," and Moholy's

fig. 7 (below)
Problems in Analysis,
Unidentified, 1958

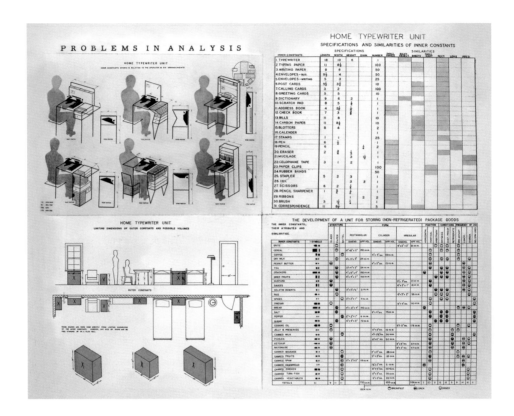

emphasis on a triumvirate of "art, science, and technology." There is evidence that we are no longer mere "problem solvers." Nor is the design profession an extension of a Madison Avenue approach to merchandising. Our work no longer constitutes a form of personal indulgence, typical of the craftsman-artist (fig. 7).

There is considerable interest today, both here and abroad, in the general study of the design process, and numerous attempts are being made to establish a methodology of design. Such attempts as have hitherto been published, however, all seem to suffer from too narrow an approach. While it may seem sufficient to study design merely from a "design point of view," such a limited attack can never adequately grasp the complexity of an object's relationship to the human environment.

It is precisely this narrowness of approach, the entrapment of the designer within a limited system, which renders the current emphasis in design education on "technical knowledge" or "craftsmanship" inadequate. Designers who proceed according to the accepted principles of their craft can produce designs that meet the requirements of those principles, but the principles themselves remain unexamined and more or less out of touch with the larger, vastly complex, and continually changing human environment that surrounds them. For the designer who is trained primarily in the material and technical aspects of his field, vital developments in the world around him—in science, business, engineering, philosophy, the arts, and so on—remain isolated from his design activities or constitute merely incidental factors. As long as he views his own field solely from the "inside" and others solely from the "outside," his understanding of his own field and of its relationship to other fields will be limited, vague, and plagued with inconsistencies. In order to overcome this handicap, the designer must leave the confinement of his isolated "subject matter" and learn to grasp the concepts that structure the design field, as well as other fields of human activity, and their interrelationships.

In practical terms this shift means that the designer can no longer be concerned solely with the development of new products or with research directed solely toward that end. Problem-bound research, useful as it is for obtaining solutions to specific current problems, must now be assimilated into a vastly more comprehensive general study of the man-made object. Freed from the limitations imposed by specific problems, such a study will result in a broader understanding of the man-made environment and, insofar as man is what he has made, of mankind itself. It is of vital importance for the advancement of design education, as a whole, that some members of the profession concern themselves with the man-made object and its environment as subjects of study at the conceptual or philosophical level. Only such a study can supply new knowledge and procedures and result in new and consistent concepts of design. Initially, the establishment of conceptual studies will affect the profession only slightly and many designers may ignore it, but after a rich body of knowledge has been accumulated, it will become an indispensable aspect of any advanced design procedure and design education.

In support of this development we may again quote the plea made by Walter Gropius some fifty years ago: "A corresponding knowledge and theory, which existed in a more vigorous era, must again be established as a basis for the practice of the visual arts." Indeed, it would be rather strange if the field of design were to be excluded from a development that has taken place in all other fields. Most areas that man has brought into his sphere of interest have historically developed in three distinct phases. First, man becomes interested in certain aspects of his environment and seeks to bring them into line with his needs or desires. He crudely manipulates certain materials or environmental factors, and in time this manipulation results in the acquisition of more refined skills. This first step in the historical sequence is practice—the direct manipulation of appropriate aspects of the environment in order to obtain certain immediately desirable results.

The field of medicine may serve as an example of the movement from this first phase to a second. It can be assumed that at one time in history man began to care for his fellow man when the latter became sick or was injured. In the case of very early man, such concern ceased with the immediate crisis that gave rise to it. Initially, any member of the group may have administered to those in need of help, but in time some individuals must have so distinguished themselves by their superior performance in various crises that the specialized practice of "medicine" came to be their recognized function in the community. The practitioner would then have had at his disposal both time and whatever resources were available to enable him to improve his skills and acquire specialized knowledge. Admission to the "profession" must have required increasing skill and specialization, until at last a prescribed program of training became the only legitimate means of entry. This process constitutes the second phase of historical development, education. At first, such training was a mere adjunct of the practice itself, a form of apprenticeship, but in time, as instruction became more and more formalized, it was separated from practice and became an institutionalized formal education.

In most fields, the development continued to still a third phase, research, which augmented the earlier amalgam of practice and education. This phase instituted a broad and general study of the concepts, methods, and materials involved in the field, a study that was independent of the urgency of specific problems requiring immediate practical attention. Finally, this phase led members of more advanced professions, no longer restricted by the limitations of specific problems, to comparative studies of concepts, methods, and materials felt to be related to their own.

Thus the full development of a professional field includes three phases:

• Practice: the direct manipulation of techniques and materials in a given field in order to obtain specific and desired results.

• Education: the preparation of individuals for practice in a given field, according to accepted principles.

• Research: the independent study of concepts, methods, and materials in a given field in order to establish an examined body of knowledge, as well as to establish its relationship to other fields.

These three phases are today typical of most recognized fields of professional endeavor, encompassing almost every area which it is useful for man to know and control. But the field which is the concern of the designer, that of the man-made object, has not progressed to the same degree, because the designer, whose concern has been the development of products, has never advanced to the stage of introducing general research. This vital aspect of man's environment has not yet obtained the benefits that such basic or general research can provide. Consequently, the practice of design remains haphazard (as witnessed by the appearance of our environment), and the core of design education is still essentially the "learn by doing" approach. While educators in other fields have been able to supplement the practical side of their enterprise with an established body of knowledge, design education remains largely a form of training. The advancement of design procedures and general knowledge regarding the object remains primarily the task of institutions of higher learning. As an integral part of design education, such study will provide a new impetus and fertile ground for the continuous development of design education and, it is to be expected, the practice of design.

The Professional Status of the Designer

In recent years there has been much discussion about whether industrial design constitutes a profession or merely a business. Many designers have volunteered a variety of definitions aimed at establishing their professional status, but only too often these statements are merely wishful thinking, having little reference to anything that might constitute the structure of a true profession. It is apparent that a profession cannot be established by merely making declarations or adopting the customs of other professions. The professional designer must concern himself with matters that go beyond the object to be designed. It is not enough that he should grow in status with each work he accomplishes. Such growth of mental and manual dexterity and proficiency only can serve him in his next task and will vanish with him. To achieve the continuous growth of the profession and, more importantly, the advancement of the cultural aspects of our visual and physical environment, the scope of the design field must be broadened. Any structural definition of the term "profession" must include the phase of general research, which assures professional status. Only this phase can provide the practitioner and the educator with a body of knowledge and a standard of performance to enable them to view specific problems in the light of broader concepts.

All of this is said in full awareness of the still prevalent notion that design is an art and not a science and that it is futile, even dangerous, to approach it in a systematic manner. That this view is gradually coming into disfavor, however, is indicated by the more frequent attempts being made to establish a methodology of design. There are those who assume that because the methodology of the past cannot be successfully applied to our present problems, all methodology should be regarded with suspicion, if it should not be rejected altogether. Consider the following quotations:

"From the very beginning there were efforts to work out formulas resolving the intuitive factors in human expression into predictable and manageable elements. There have been repeated efforts to lay down canons which, if followed, will assure a harmonious attainment in a chosen field of creation. We have become skeptical about precepts of harmony constructed in this way. We no longer believe in mechanically producible works of art." [23]

"YOUR PROFESSION IS UNIQUE—KEEP IT THAT WAY...There has been a tendency on the part of some in the industrial design profession to try and reduce the practice to a science... It has nothing to do with science."[24]

"The practice of design isn't going to change into anything other than what it always has been, or should have been..."[25]

"Do not create a Frankenstein of science or technology out of a fundamentally humanistic art."[26]

Of course, it is possible that these concerns have now been set aside. Nevertheless, these statements constitute an historical account of a phase in the development of the design profession.

Some designers may feel that the particular manner in which they are introduced to the creative process, their particular working habits, and the fact that the field does not entail a body of knowledge are unique characteristics of their profession. They do not realize that these are only phases in the development of the design profession. Perhaps the great innovators of the recent past, whether in science, engineering, architecture, art, or specifically in the field of industrial design, are still too vivid in our minds; and for this reason we are tempted to follow the pattern already laid down in educational development, without realizing the drastically different situation we are confronted with now and will be confronted with in the future.

The fact is that the profession and the educational establishment of today provide little or no direction for the development of design practice and education, as the Bauhaus once did. The question is, what will be the future of the design profession? The prime task of the professional designer has been to develop new products and, only as a matter of individual inclination and experience, to concern himself with the advancement of his profession. In the future, the profession of designing must include within its ranks members whose prime concern is the study of man-made objects as a formal discipline independent of specific design problems. In support of such a concern, the late Peter Muller-Munk, FASID, past president of the International Council of Societies of Industrial Design (ICSID), made the following observation in an address before the second ICSID general assembly in September 1961:

In order for an occupation to become a profession it must accumulate a body of learning and a recognized set of educational disciplines which can be conveyed to our successors, to the young students and apprentices who must carry on from where we leave off. It is not enough for a few patriarchal leaders to share their secrets or to exchange their experiences in the privacy of their special exclusive little clubs. Without an educational system to give continuity to the practice of a profession, and without a set of knowledge basic for professional performance, there is no profession. A profession must develop its own international alphabet, which it must teach to others, so that the language of the profession cannot get lost. The torch must be handed on, or else it expires forever. [27]

The notion of a theory of design may seem a disreputable one to some practitioners in the field. This is not without reason, since too often in the past the designer has considered mere verbalization as theory. Serious professionals are bound to be wary of such an imposition. A theory is like a map—an invaluable tool when it corresponds to reality; worse than nothing when it is false to reality. Nevertheless, the need for knowledgeable designers has been voiced frequently in recent years. However, a concrete proposal with regard to the attributes and education of such a person is still lacking. One thing is certain: the knowledgeable designer will not be brought into being by a declaration. The task of building up a body of knowledge concerning the man-made object will be as laborious as in any field. The use of such an established body of knowledge will not be altogether a matter of personal choice. At the present time the designer can still maintain a "take it or leave it" attitude and proceed in the conventional manner, relying essentially on his native abilities. But there are strong indications that drastic changes are taking place in the field, changes that seem to follow the pattern of development in other fields. As soon as a substantial body of knowledge has been accumulated in the field of design, it will be foolhardy to ignore it.

An important point to remember is that we are increasingly aiming for an integrated environment. While our man-made environment is, and will remain, the product of the collective efforts of numerous individuals, each following personal inclination, the results of these collective actions will be experienced by the users as a whole. Because of his particular activity, the designer concentrates directly upon a single unit. The user, however, experiences a unit not merely as an aspect of a complex physical environment, but as a part of his daily routine of living. Thus the user approaches an object from a different point of view. This fact should have a bearing upon the development of products.

Earlier societies were able to achieve relatively integrated environments essentially because of the dictates of a few and the indifference of the remaining many. Such conditions are no longer acceptable or practicable. Our present mass-producing, democratic society can achieve such integration only through the establishment of new and broad formulations that result in a body of knowledge.

The lack of a comprehensive theory of design should not be attributed to any failing on the part of the design profession. There are other factors at work in the field itself that are responsible for the relatively slow and late development of design theory. It is true that man has utilized and developed objects for countless ages, but it is also true that the design field comprises what is perhaps the most diversified pursuit of man, encompassing as it does practically all aspects of life. This suggests an approach quite different from that of the Bauhaus with its educational emphasis on skill (craftsmanship) and native intuition. Nevertheless, the approach I suggest is a logical continuation, not so much because I am a product of the Bauhaus, but because I see in the early Bauhaus approach the origin of the general ideas expressed here. The Bauhaus accepted the machine as a partner in the task of fulfilling the needs of our society. We must now, as it were, accept the intellect as a worthy partner in design. Leonard da Vinci wrote: "Those who are enamoured of practice without science are like pilots of ships without rudders or compasses, who are never certain where they are going." [28] And Josef Albers described, "I believe that thinking is necessary in art as everywhere else and that a clear head is never in the way of genuine feeling..." [29]

2

The Evolving Man-Made Environment

A biological entity does not exist by itself or for itself. It retains its integrity through interaction with other entities. This is not only true for plants and animals but for mankind as well. In this bondage, early man spent much of his time merely sustaining life by procuring food and securing shelter and safety.

If we describe man's early history in a conjectural and schematic way, we can assume that he made use of available environmental means directly and spontaneously, gathering food for immediate use and seeking shelter when conditions required it. He thus obtained a direct benefit. Two phases were involved in this process: the gathering of means and their utilization. It is assumed that initially these two phases were more or less continuous. In time, a separation took place. Gathered food was not necessarily consumed immediately but was stored for future use. Hence, a third preparatory step was introduced—that of storing—which can be referred to as the beginning of a form of "anticipation" and "deliberation." This preparatory step required, at least intermittently, a certain freedom from want and a temporary release from the concerns of subsistence.

The act of storing permitted a greater and better selection of gathered means, which gave man a certain measure of control over his needs. This initiated further development: the modification of gathered means for the purpose of serving those needs more appropriately.

At this stage in man's development, the process of sustaining life involved four phases: gathering, storing, modification, and utilization of essential means. Whereas the storing of food required only a place for storage (passive means), the modification required the use of other (active) means. In early times, these other means were the bare hands and teeth used to modify an entity in order to prepare it for utilization. Progressively these were supplemented and ultimately replaced by other available means–hence, the advent of the tool.

The Tool

Along with the gathering of food, man began to select certain entities for the sole purpose of obtaining and/or modifying the means essential to his sustenance. The benefit derived from the use of these tools was purely indirect. Throwing a rock (a tool) to kill an animal, for instance,

fig. 8 (below)
A Woodlathe,
W. Watkins, 1958

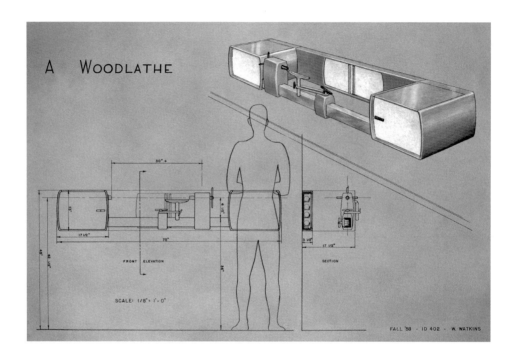

was an indirect use of an entity, which in turn obtained a direct benefit, the provision of food.

The Advent of the Man-Made Object

The use of such existing means as a stone or a stick also is common among certain animals.
Some not only make use of such available environmental means as tools, but in certain instances
will modify these natural products to serve a particular task better. For instance, an ape may
break a twig from a bush (selection), strip the leaves off (modification), and insert the twig
in an ant hole. Thus the twig is used as a tool to facilitate the getting of ants from a hole in the
ground. Or he may use a stick as a tool to reach a banana, provided that such a tool is at
hand. The ape uses these means indirectly to obtain something which will serve him directly.
Most likely, he will discard the stick once it has been used, making it once again a mere
environmental detail. The stick is only a "tool" so long as the animal uses it. However, man

advanced beyond this stage when he (1) selected, used, and retained a means for possible future use, and (2) selected, modified, and used a means for the sole purpose of modifying another means. Thus he uses a tool to make another tool, which in turn may be used to make yet another tool, and so on.

While some animals may use a tool to help obtain a direct benefit (e.g., food), man has the ability to improve the process of facilitation. Only man makes a tool for the sole purpose of making another, which after a series of such processes ultimately benefits man directly (fig. 8).

The Two Worlds of Man

Before the advent of the man-made object, when man was merely part of the ecology of a natural environment, he had to cope with only one world—nature at large. However, through millennia of continuous action, man created a new environment for himself. Although existing within nature at large, it is essentially a different kind of world. Now man has to cope with two environments—the natural and the man made.

Historically, the development of the man-made world began with the introduction of artifacts. Prior to this time, matter (food, protection, etc.) was used only to obtain direct and immediate satisfaction. However, when man also began to use available matter in an indirect way— i.e., to obtain not immediate but delayed satisfaction—he initiated a development which has continued into, and will continue beyond, our own time. From only a few artifacts which merely assisted in daily tasks, the man-made environment has grown to engulf modern man and has become essential for our very survival.

Man's Relationship to the Natural and the Man-made Environment

While most of us today depend almost entirely on our man-made world, we still have to maintain a direct relationship to the natural environment, not only as a source of enjoyment but also to obtain raw materials and establish new frontiers. In order for man to operate in a natural environment, he must adjust to the situation at hand. Nature is not designed to accommodate man. He has to cope with it on nature's terms; to climb a mountain, for instance, man has to

fig. 9 (opp. page)
A Drill Press,
J. W. Goodson, 1959

use whatever "steps" are available. The success of such a venture depends entirely on the ability to adjust to a particular situation. This unilateral adjustment by man constitutes an asymmetrical relationship.

Quite a different relationship exists between man and his man-made environment. One may hastily conclude that this, too, is an asymmetrical relationship. However, it is the reverse of man's relationship to nature in that every effort is made during the design procedure to adjust a unit to its user (fig. 9). One could thus assume that the object is being adjusted to man and not the reverse. However, this is not the case. Although the designer makes every effort to adjust the object to the user, the user still has to adjust to the object when he uses it. We are dealing here with a symmetrical relationship. This can be illustrated again by someone ascending a height, but in this case by means of a man-made stairway. Stairs are designed to fit man so far as the height of the riser, the depth of the step, etc., are concerned. But the user still has to adjust him to the stairs. This is because man is not naturally or instinctively endowed to use an artifact. He has to experience, to learn, or to "get used to" it. This may not be obvious in using a stair- way or a chair, but it becomes quite clear in using certain tools or a bicycle, driving a car, or flying an airplane.

The user not only adjusts himself to the artifact but in doing so undergoes certain changes. He acquires skills and some knowledge of the object's operational effectiveness. These exper- iences in turn affect the attitude of the user, not only towards the object concerned but, in more general terms, towards other aspects of the environment. Put differently, after using an artifact, the user is no longer the person he was before using it. Changes may not be discernable on an individual basis, but they become obvious when we view the historical development of man, particularly since the industrial era. The driving of an automobile is a typical example of such a change. At the beginning of this century, a speed of seventy to eighty miles per hour would have been considered that of a daredevil, but today it is common and generally accepted. Not only have our roads and cars vastly improved, but today's drivers, and non-drivers, are not the same as they were around the turn of the century. Skill, knowledge, and attitudes have changed and, I might add, will continue to change.

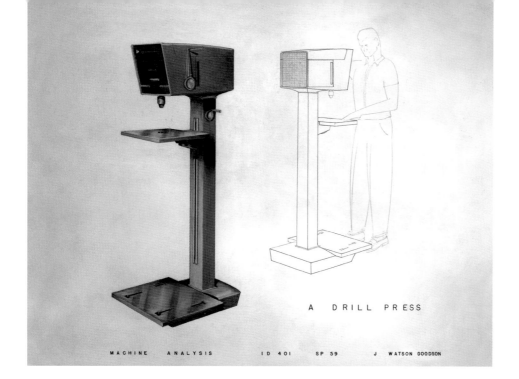

A   DRILL   PRESS

MACHINE   ANALYSIS   I D 4 0 I   SP 59   J   WATSON GOODSON

Yet regardless of the ever-growing, man-made environment, nature remains our host and provider. Although nature harbors mankind and seems to be hospitable, it also is hostile and demanding at times. The underlying tendency of much of man's effort is to transform the hostile aspect of our environment into a hospitable one and thereby gain some measure of control. Although the history of man indicates the achievement of ever-greater control over ever-wider aspects of our environment, the limits of this control are readily apparent. Not only do such dramatic events as earthquakes, tornadoes, and floods, but also many daily and often minor events affecting only individuals remind us of these limits. All of these occurrences represent a challenge for those who are concerned with increasing man's control.

Classification of the Man-Made Object

In the preceding section, we have dealt with the two environments of man—the natural and the man made—and man's position within and relationship to these environments. We now turn to the man-made object itself.

Although the man-made object exists in a space-time continuum, we will consider the spatial and temporal aspects separately, in accordance with the following outline:

1. Spatial: the study of the spatial aspect includes three sections.

a. Classification: the systematic grouping of the man-made object

b. Attributes: the basic characteristics of the man-made object

c. Transformation: transforming matter to bring the man-made object into being

fig. 10 (opp. page)
Furniture, W. E. Lerdon, 1958

2. Temporal: the study of the spatial aspect includes two sections.

a. The general development of the man-made object over a period of time

b. The development of specific objects

The following classification is an initial attempt to establish a systematic grouping of man-made objects. While much remains to be done before a satisfactory classification can be obtained, in its present form, it already provides some assistance to the designer. Further, it may provide an impetus for others to improve upon this beginning.

In most areas in which man has ventured to investigate, he has established a consistent classification of the matter with which he is dealing. This is true, among other fields, including botany, zoology, and chemistry. Some of these fields actually have two classifications: a popular one, which is part of the spoken language, and a scientific one, deliberately developed to serve a systematic investigation of a particular field. While a popular classification is sufficient in our daily life, the scientist requires a consistent classification to facilitate research. The progress made in these fields is partly due to such a systematic classification.

In the area of the man-made object, we find only one classification—a popular one used by laymen and professionals alike. The man-made object is classified in such vague groupings as tools, appliances, utensils, furniture, flatware, and numerous others (fig. 10). No consistent system underlies these groupings; and terms are often used interchangeably. While this identification of a great variety of objects may serve in our day-to-day activities, it provides little or no assistance to the professional designer in the development of new products.

In scientific fields, classification is based on features and specific characteristics. This is called a "natural" classification. A comparable classification is not feasible for the man-made object. All tables, for instance, belong to the same group regardless of how many legs they may have. A classification of the man-made object only can be based on the usage (or purpose) of an object. This is called an "artificial" classification.

What are the major groups of man-made objects? Since classification is based on usage or purpose, the question is, in what different ways objects or any entity can be used? It may be

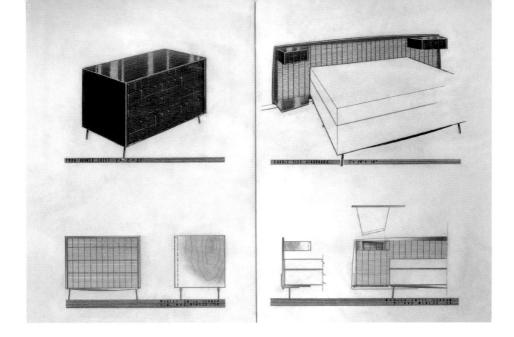

surprising in view of the vast number of existing man-made objects that there are basically only three ways to use an object. In order to deal appropriately with this aspect of our classification, it will be necessary to refer not only to actual man-made objects but to matter or entities in general. This is essential, for it would be difficult to determine at what point a certain means constitutes an object. A rock, for instance, can serve as a substitute for a man-made stool. We must thus ask the question; in what different ways can physical entities be used by man?

• A physical entity can be used to transform other entities; an object designed to transform other entities will thus be referred to as a *transformer type of object*. Such an object is used to effect changes in other objects, environmental aspects, or in man himself. It represents the largest group of man-made objects and includes, among others, tools, appliances, furniture, cars, and houses.

• A physical entity can be used to convey a message or meaning; an object designed for this purpose will be referred to as a *symbol*. Such an object has had a message or meaning arbitrarily attached to it. The group includes, among others, printed matter, paintings, sculpture, and clocks.

• A physical entity can be merely manipulated; an object designed for such manipulation will be referred to as a *leisure type of object*. Any item used by an individual or group at play, such as various kinds balls, tennis rackets, golf clubs, etc., belong to this group.

A general remark is in order before we consider each of these groups further. Since our classification is based on usage, any object can serve on occasion in any of the three categories. A chair, for instance, though designed for a specific purpose, can nevertheless be used as a symbol; provided, of course, that the essential prerequisites for use as a symbol are met. This does not mean, however, that all objects (as is often suggested) have symbolic qualities.

The characteristics of a symbol are discussed below. If a chair were used in a balancing act, it would belong to the leisure group. Such exceptions do not invalidate the classification and will be considered further in a later chapter.

The Transformer-Type Object

Strictly speaking, any object that is used to obtain certain desired changes in other objects, environmental aspects, or in man himself, can be considered a transformer object. The physical use of the object is the decisive factor. However, our concern here is exclusively with those objects which are specifically designed for such a purpose. There are two kinds of changes that man seeks to bring about. These correspond to the two types of transformer objects: the defensive type and the offensive type, which are defined as follows:

1. Defensive Type: the defensive object prevents certain changes from occurring in order to maintain the status quo of certain environmental aspects. To understand the kind of change involved, one must realize that our physical environment is in a constant state of flux. Some changes are rather drastic, even catastrophic, while others are imperceptible. Some are beneficial to man; others are detrimental in varying degrees. The defensive object seeks to intercede and divert an undesirable change. It serves as a kind of shield, which partly (i.e., selectively) or completely blocks or diverts impinging forces. It does so passively or defensively; that is to say, it does not change the impingement itself. It is a kind of container or holder, which effects changes by preventing or delaying something that would otherwise occur.

The group of defensive objects has two sub-groupings: selectors and holders.

a. Selectors: a selector serves to prevent certain changes caused by natural conditions, such as precipitation, temperature, gravitation, radiation, fog, and dust, or man-induced conditions, such as vandalism, stealing, and defacing. The selector seeks to maintain the physical integrity of other objects. Cabinets, houses, refrigerators, and umbrellas belong to this group. A house, for example, is selective in that it blocks out wind, rain, dust, and noise but lets in light; an umbrella shields from rain but not from cold.

b. Holders: a holder serves to maintain the positional integrity of an entity, counteracting gravitational and other physical forces. Tables and chairs are typical holders. A simple hook, coat hanger, or bowl would also belong to this group. While some objects are obviously either selectors or holders, others combine both characteristics. We will call these selector-holders. An example of this type of object would be a milk bottle, which prevents (actually only delays) changes due to atmospheric and gravitational conditions.

The basic characteristic of a defensive object is that it not only contains something but is at the same time itself contained (by other objects and/or the environment). Thus it has a twofold relationship: to that which it contains and to that which contains it. This twofold relationship is an important factor in designing a defensive type of object. In a design procedure, that which the object contains will be referred to as the "inner constants," and that which contains it, as the "outer constants." The term "constant" signifies that these have to be considered as they are; generally they cannot be changed. There may be exceptions, so far as the inner constants are concerned. If the procedure has been extended to include the designing of the inner constants, these exceptions will be referred to as "variable constants."

Whereas the inner constants in a design procedure are determined by the kind of object being designed, there are three kinds of outer constants which may have some bearing upon the object itself. These are 1) other objects, 2) prevailing conditions, and 3) man or the user.

Any or all of these will have a particular relationship to the object being designed. More will be said about this relationship in the section on design in Chapter 4.

2. Offensive Type: an offensive object is used to change the physical characteristics of other objects, persons, or environmental aspects. Whereas the defensive object is used to prevent changes, the offensive object is used to induce changes. Strictly speaking, the defensive object is what we call a tool, but a spoon, a pencil, or even a rock or a branch from a tree may qualify—in short, any object or means used to transform the physical characteristics of another entity.

Our emphasis will be on objects specifically designed to serve as tools, which are intended to change one or a combination of the basic attributes of another entity—structure, form, or position.

fig. 11 (opp. page)
Electric Wall Clock,
J. Mengason, 1959

The Symbol-Type Object

A symbol type of object is used to convey a message or meaning. Such a message is not inherent in the object's physical characteristics but has been arbitrarily attached to it. For example, the symbolic aspect of the letter "A" does not refers to the physical characteristics of print, ink, paper, reflected light, or any other observable aspects. These serve only as a medium for conveying the meaning "A." In fact, when we look at a symbol, we should remember:

• Physical characteristics are being ignored in favor of a superimposed message (although our interest may occasionally be directed precisely to the symbol's physical characteristics); and

• Since a symbol does not represent itself (as, for example, a transformer object does), it has to be interpreted in order to be understood; the message is in code form and has to be decoded or interpreted.

The interpretation of a symbol has to be based on a prior agreement between the sender of the message and the receiver. Without such prior agreement, a symbol cannot be understood. This is the basic characteristic of all symbols. There are times when one may understand a symbol without being aware of a prior agreement. But an understanding still exists; it is based on a tacit agreement reached between persons (usually a small group) who experience a certain event in common. An object that may be part of such an event, directly or indirectly, could assume the role of a symbol. It may have merely souvenir qualities, with meaning only for some, and as such it could gain wider acceptance through negotiation. The vast majority of our symbols today are based on negotiated agreements deliberately arrived at. These represent our common means of communication.

Types of Symbols

There are four types of objects in the symbol group, each serving a different purpose:

• Transmitters: a transmitter symbol conveys a message for the purpose of interchanging thoughts, feelings, and concepts. Our means of communication, language, writing, illustrations, paintings, and sculpture, belong to this group. Traffic lights, street signs, directional arrows,

ELECTRIC WALL CLOCK   JAMES MENGASON   I.D. 303   SPRING 1959

to name only a few, similarly serve to transmit a more or less specific message to others.

• Indicators: an indicator symbol indicates the presence, nature, intensity, magnitude, rate of change, and similar aspects of certain physical phenomena. Its operational principle is based either on response to environmental occurrences (for instance, a barometer, which responds to air pressure, or a thermometer, which responds to temperature), or on an established event that parallels other events (for instance, a clock or a ruler). In this capacity it allows for greater objectivity in communication (fig.11).

• Models: a model symbol represents a specific concept. It serves to vivify that concept and make it understandable. A designer's sketch, drawing, or mock-up, a chemist's molecular model, a biologist's genetic helix model, an economist's diagram or chart, or a geographer's map are all model symbols. Although usually conceived by a designer or scientist for the purpose of understanding a newly developed concept, such an object becomes an important general means of communication.

• Manipulators: a manipulator symbol manipulates data and information. Included in this group are such instruments or machines as a calculator and a computer.

The Physical Aspects of a Symbol

Since a symbol object is a physical entity, it has characteristics which are similar to those of the transformer object. Although the physical aspects of a symbol are secondary to its message, they are, of course, important in the design process. A traffic light on a street corner is obviously a symbol object conveying a message. Its symbolic features are its alternately flashing green, red, and amber lights. In order for these to function, a variety of physical components, such as lamps, lamp sockets, switches, and relays, are required. They, in turn, have to be held together and protected from prevailing conditions. The various components are all similar to those of transformer objects. However, in a symbol object, the physical components are subordinate to the message being conveyed, whereas in the transformer object, the physical components are themselves of prime importance.

There are objects that combine certain characteristics of both the transformer and the symbol group. These are simultaneously utilitarian objects and conveyers of messages. In point of fact, a symbolic meaning can be attached to any object—for example, a bishop's miter—thus making it serve a twofold purpose. The miter functions as a utilitarian cover for the head. But this function has also been made to symbolize a rank and position in certain segments of our society. The utilitarian aspect may not be important, but it is nevertheless present. Relatively few transformer objects are also symbols. When this is the case, such objects usually (perhaps always) serve certain ceremonial and/or ranking purposes. Generally a transformer object is modified as a symbol; rarely is a symbol object modified to serve a utilitarian purpose.

The Leisure-Type Object

When a transformer object is used, the whole action is directed towards obtaining certain desired results, which in one form or another, affects the life of the user. This is the sole reason for having and using this kind of object.

In the overall development of transformer-type products, the general tendency has been to simplify the process and obtain increasingly faster results. Generally speaking, one does not "indulge" in the process. This also applies, to some extent, to a symbol object; the tendency has been to convey the message with increasing quickness and precision. In both cases— ignoring any frivolous handling—the aim is to obtain the end-result as fast as possible. Numerous cases can be cited to illustrate this tendency. One known to everyone is the "instant" product offered in our contemporary grocery store.

The leisure object serves a very different purpose. Here the process of manipulation is the main reason for its existence. The user indulges in the process. In using such an object, one does not seek to effect environmental changes. Arbitrary rules and regulations often govern its use.

Again, any kind of object may at times be used purely for manipulation. We will refer here, however, to objects solely developed for this purpose.

3

The Basic Attibutes of Objects

Every object or entity, whether natural or man made, a mere rock, a piece of wood, a chair, car, or house, has three basic characteristics or attributes:

- Structure: the inner build-up of the object;

- Form: the boundary of the structure; and

- Position: the physical relationship of the structure-form to other structure-forms.

These three attributes represent the integrity of all objects. No object exists without having a structure that is spatially restricted by a boundary, and every structure-form is positioned in relation to other structure-forms. In a design procedure, the designer determines the specific characteristics of each of these attributes.

Form plays an important role in our environment, and it is for this reason that the designer is sometimes referred to as the "form-giver." However, one must always keep in mind that a form cannot exist without its two companion attributes: inner structure and position. Although some designers may consider the form of objects their only concern, an object also has structural and positional attributes. By neglecting to determine, or unwittingly ignoring, the specific characteristics of structure and position, the designer consigns these attributes, incorrectly, to the role of prevailing circumstances. An object is always a complete entity—i.e., it always has structure, form, and position.

The Attribute Structure

Structure, the inner build-up of an object, is basic not only to every monolithic unit but to every group or assemblage of single or multiple units—for example, the arrangement of furniture in a room or the layout of streets and buildings in a city. The term "structure" will, therefore, be used rather broadly in this context. We will define it as: whatever is within the confines of a considered whole. The whole represents the form. It makes no difference whether the unit or group is natural or man made, whether it has been designed or is the result of natural or accidental circumstances. This implies, of course, that every monolithic unit is being considered both as a whole and, from a wider point of view, as a mere detail of a larger whole. The larger whole, in turn, can be considered as a mere detail of an even larger whole, and so on. In fact, structure permeates our total environment, whereas form represents merely an aspect of the total environment.

It would be desirable to establish a comprehensive classification of all existing and possible structures, from the minutest particle to the most monumental man-made structure. However, this is a task for the future. The following limited grouping may nonetheless be of some use to the designer. It should be considered as tentative and open to further refinement, delineation, and extension. The groups classify the different kinds of matter which can exist within the confinement of a form of any figuration. For simplification, the reader may assume the form of a cube. The inner confines of such a cube could consist of any of the following structures. Most objects, however, consist of various combinations of different structures.

1. Static structures: units that have a rigid form.

a. Compositional structures: the inside of the form consists entirely of a man-made or natural material, such as a rock or a block of wood.

b. Built-up structures: structures consisting of a number of parts. The inner build-up of the form may consist of the following:
- Lamination: layers of sheet material (plywood, etc.)
- Modules: regular or irregular pieces (bricks, gravel, etc.)

c. Frame structures: the inner confines of a form may consist of the following structures made of sheet (shell) and/or linear material:
- Shell-contour: a shell following the contour of a form
- Shell-space: a shell within the inner space confined by a form
- Skeleton-contour: linear parts following the contour of a form
- Skeleton-space: linear parts within the inner space of a form

2. Systems structures: the inner structure of a form consists of interrelated mechanical, electronic or chemical operational units, such as a watch or a toaster.

3. Variable structures: units that have a structure which allows the changing of the form.

a. Adjustable structures

- Multi-purpose: serving two or more usages; a step-stool, etc.
- Units adjustable to different outer constants, such as:
  - The stature of a person (anthropometric)
  - Various environmental aspects (other objects, etc.)
  - Various conditions (indoor, outdoor, etc.)

b. Access structures: structure providing access to an enclosure.

- Shell opening: a part of the enclosing shell opens
- Split opening: the enclosure splits at some point (refrigerator)

4. Dynamic structures: the structure-form is maintained by an inner force: air (balloon or tire), fluid (waterbed), granular (sandbag), spring (upholstery); it is maintained by an outer force: gyroscope, spinning top.

5. Operational structures: whereas all the preceding structures involve single units, the operational structure involves a number of such units. The fact is that an object becomes operational (usable) only in conjunction with other objects, environmental aspects, and man. The operational structure refers to the minimum number of objects, environmental aspects, and persons required to obtain a desired result, for instance when slicing bread or typing a letter. This definition is rather vague and would be even more so if we were to consider the larger structures of which the operational structure is merely a part. A more precise definition will have to be formulated in the future.

The Attribute Form

Form represents the boundary of a structure viewed from the inside. When the same form is viewed from the outside, it represents an aspect of the environment which is under consideration for some reason; either because reference is being made to it, or because it is being used or dealt with in some way. It may be a monolithic unit, a part, a single object, or a number of units (several parts of an object or a group of objects, an arrangement of furniture in a room, or the layout of a city). Form is a "relational" quality in that its configuration is the result of or is

fig. 12 (opp. page)
A Display Unit,
W. G. Watkins, 1959

determined by its relationship to other forms existing in our environment, including man. This relationship is twofold: to certain manufacturing or natural processes and to certain utilization processes.

It would be to our advantage to establish some order in the vast variety of forms in our man-made environment. But, it seems to be, the very diversity of these forms makes a meaningful classification impossible.

Surface

The surface of an object represents a sub-attribute of form. All surfaces have two basic charact-eristics: color and texture. A surface may be black, white, or any other color; and it may be smooth, like the surface of glass, or rough, like the bark of a tree. Color refers to the reflection of light, and texture to the peaks and valleys of a surface. The texture of a material may be minimal, as in glass, and special instruments may be required to measure them. No matter how minimal, all surfaces have texture. It is important to approach surfaces from this broad perspective. Otherwise we would have considerable trouble deciding at what point a given surface can be said to have texture.

In the past, more so than today, the designer of an object primarily concerned himself with its form. The surface texture was often an incidental by-product of the manufacturing process, and the color a by-product of the process of preserving the object. The deliberate use of texture as a surface treatment is a recent innovation. Textiles and, perhaps, ceramic products are an exception, since texture was often a major aspect of their design. Since color has been dealt with extensively in special books and educational courses, the emphasis here will essentially be on texture.

Texture is now receiving increasing consideration from designers. A number of events in the recent past may have contributed to the designer's awareness of texture. In the field of fine arts, for instance, the futurist F.T. Marinetti published a paper after the First World War about "tactilism," which referred to the tactile sensation of material and the possibility of creating a "tactile symphony" by moving the finger over a pattern of different textures arranged to create

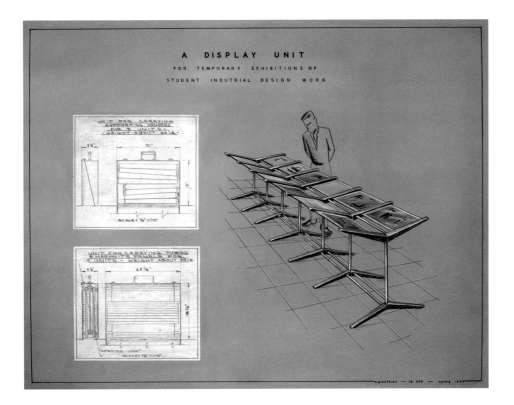

a pleasant tactile sensation.[30] When Moholy-Nagy, continuing Marinetti's idea, introduced the exercise referred to as a "tactile chart" at the Bauhaus in the 1920s, further attention was drawn to surface texture.

A number of other events, which appear to be unrelated to the field of design, also drew attention to texture. One such event occurred in the field of lighting. Throughout history, a prime concern of man has been to obtain a reliable and economical source of light. With the development of the incandescent bulb, we gained an abundance of light. This shifted the concern toward improvement of the quality of lighting and to appropriate light fixtures. Originally, a light fixture merely served to hold a source of light (a chandelier, for example). However, since the quality of light in a given environment is not solely a matter of an appropriate fixture but also depends on the color and texture of walls and ceiling, a proper approach to adequate lighting in a room required an appropriate treatment of its surfaces.

Industry also contributed to a new interest in texture by creating a number of composition boards: Masonite, Celotex, Homasote, to name only a few (fig. 12). Technology and commercial specifications required the determination of two variables in these products: size and texture. The former resulted in the familiar 4' by 8' panel and directly related dimensions. (Although such measurements often influence design, they will not be discussed at this point.) The surface texture of these composition boards initially resulted from the manufacturing process (pressing the material between two smooth platens). In time, some boards were deliberately textured to imitate other materials (such as leather or woven cane). These textured boards demonstrated the potential inherent in the manufacturing process; and this eventually resulted in the availability of a variety of textured panels made of plastic, metal, plywood, and ceramic materials.

The precision machine industry developed not only a terminology of surface textures, but also a kind of sample chart showing different textures. This was to be used for comparisons and to judge certain surfaces. Had Moholy lived to see this development, he would undoubtedly have been delighted. Besides such comparison charts, instruments were developed to measure surface roughness. It should be pointed out that such "charts" and "instruments" are used only in particular sectors of our industry. No similar concepts involving consumer products are available to the designers. Essentially they have at their disposal only a vague terminology; in the case of painted surfaces, for instance, such terms as *gloss, matte,* or *eggshell.* The term *textured* used in reference to a textured surface is itself too vague. Contrast this to the terminology available for color. Here we have definite systems of identification, for instance, the Oswald system. These permit specification of particular colors and the certainty that what is specified will be delivered. No comparable system exists for texture; and yet, any material having color must also have texture, and as is well known, texture will affect the appearance of color.

It is evident that the establishment of a system of both classification and terminology for surface texture and texture patterns would serve to improve communication. It also would facilitate the creative effort of determining the finish of a product. The following proposal for such a classification approaches the task from a rather broad point of view, encompassing all textures and texture patterns, regardless of their practical application.

Basic Surface Textures (classified on the basis of origin):

• Natural surface: the result of a natural process: the bark of a tree, human skin, the surface of a seashell, or sand on a beach.

• Formed surface: the result of man's modification of the form of a material. It represents a by-result of a process: sawed, split, or broken surfaces.

• Finished surface: the result of a deliberate modification of a surface: planed, sanded, ground, polished, hammered, etched, indented, or perforated surfaces.

• Added surface: the result of superimposing other materials on a surface: painted, plated, or papered surfaces.

• Composed surface: resulting from the production of a specific material: a textile, ceramic, felt, paper, or composition board.

Basic Surface Patterns are classified on the basis of characteristics and arrangement of surface elements resulting in a pattern. Surface elements can have any shape, from mere dots and lines to flowers or some other representation. The following surface patterns are deliberately designed and are related to the last three categories listed above.

Basic Surface Patterns:

• Omni pattern: a uniform texture without recognizable surface elements: paper, concrete, metal, etc. These constitute the ground or base texture for the following arranged patterns.

• Mono pattern: a one-directional arrangement of surface elements.
If extended, it retains its character.

• Dual pattern: a two-directional arrangement of surface elements.
If extended, it retains its character.

• Multi pattern: a multi-directional arrangement of surface elements.
If extended, it retains its character.

• Scattered pattern: an arrangement of surface elements resulting from the application of a force: spattered paint, tossed granular materials, or a cracked window.
If extended, a different pattern will result.

• Self-contained pattern: any arrangement of surface elements that results in a different pattern when extended: any shape, or abstract or representational form.

• Random pattern: any arrangement of surface elements that is the result of an uncontrolled process.

This classification can only be considered as an initial step toward establishing a system serving a similar purpose as the various existing color systems. Its limited, perhaps tentative, listing can aid communication and enable designers to specify more precisely the surfaces of their products. Beyond this it can facilitate the creative process by enabling the designer to oversee available potentials.

The Attribute Position

Position relates a structure-form to other structure-forms. It may not be obvious that an object's position constitutes an essential attribute, such as structure and form, which is determined in a design procedure. However, a chair lying on its back is as unfinished as one structurally missing a leg. It requires a certain operation in order to become useful.

Of the three attributes, position is the least talked about, although it is present in every design procedure. One reason for this neglect is that the position of an object is simply taken for granted. Another is that we are what may be called object-oriented. We concentrate on objects more than on relationships. Yet an object (structure-form) only becomes useful when it is given a position relative to other objects, the environment, and man.

Three factors determine the position of an object: its usage, its physical contact with other objects and environmental aspects, and its arrangement relative to other objects.

The Different Positions of an Object:

1. Positions related to the usage of the object:

a. Operational position: the position in which an object is used, e.g. a knife slicing bread.

b. Ready position: the dormant position of an object that is ready for use, e.g., a knife on a table ready for use.

c. Store position: the dormant position of an object in between being used, e.g., a knife in a drawer

d. Transit position: the position of an object being moved from one place to another.

e. Display position: the position of an object being inspected, e.g., in a store or show window.

f. Prepare position: the position of an object itself being readied, maintained, or disposed of.

g. Default position: the arbitrary position of an object when the above positions are disregarded.

2. Physical contact of an object with other objects and environmental aspects (contact relationships are expressed in terms of "fits").

a. Form-fit: when the form of an object, or part of it, conforms in some way to the form of another, e.g., the handle of an object to the hand of a person, or a cabinet to a wall and a floor.

b. Variable form-fit: when the form of an object, or part of it, conforms in various positions to that of another, e.g., the handle of a screwdriver.

c. Movement-fit: when a moving or sliding object conforms in its parts to another object, e.g., a sliding door, or a car with the bank of a road at a turn.

d. Space-fit: the spatial relationship of an object to other objects that are not indirect contact with it, e.g., a person in a narrow hallway or telephone booth.

e. Default-fit: the uncontrolled relationship of one object to another when they are placed at random.

3. The arrangement of two or more objects according to:

a. their physical characteristics of structure and/or form and position

b. the environmental space they occupy

c. The preference of the user: this facilitates the recognition of the units involved, their order and frequency of handling, and the manner in which they are manipulated

What is the importance of being aware of the different positions of an object—in other words, of its "twenty four-hour" aspect? In general, the prime emphasis in a design procedure is on the operational aspect, the function, of the object being designed. Although the usefulness or operational aspect of an object is primary, the object also has an existence when it is not in use.

The question arises, what are the characteristics of an object when it is not in use? The above listing of potential positions draws attention to this aspect.

4

The Transformation Process

Now that we have dealt with the various kinds of man-made objects and their basic attributes, the question must be asked, how do these come into being? This takes place in a transformation process, which transforms matter. The process includes three sectors:

- Matter: this sector represents the material or entity which is being transformed.

- Forming: this sector represents the means which transforms matter.

- Design: this sector directs the forming process which transforms matter.

These three sectors are involved not only in the development of a product but in any action by man that results in environmental changes. Thus the mere cleaning of a room or slicing of a loaf of bread involves a transformation of matter, a certain means for effecting change, and a design decision that directs the process. We are referring here to only one transformation process, but most of our products are the result of a number of such processes following each other in succession. The result of one transformation process becomes the matter to be transformed in the next stage. Generally, each stage in this complex process is directed by a different profession. If we outline in brief the process involved in the development of a product, we might begin, for instance, with the prospector or geologist who selects raw material from nature. After its transformation, this material becomes the matter to be transformed in the next process, which is directed by the technologist or chemist, employing an altogether different forming process. The results are in turn transformed further into parts, preforms, and components under the direction of different kinds of engineers. Ultimately, their results are converted, under the direction of engineers, architects, and designers, into the vast variety of products that make up our man-made environment. We will consider each of these sectors in detail, beginning with the matter sector.

The Matter Sector of the Transformation Process

The matter sector refers to anything subject to transformation. It may be a single unit, such as a raw material; a preformed part or a group of entities, such as an arrangement of furniture; or part of the environment. The general characteristics of this matter are the basic attributes of structure, form, and position, which have been dealt with in the preceding chapter. One, or any combination, of these attributes may be subject to transformation.

The following rather brief listing merely serves to indicate the range and variety of the matter that can be the subject of a transformation process. Each category is itself the result of a transformation process.

- Raw material: found in nature or as a by-product of a manufacturing process.

- Selected material: ore, rock, lumber, etc.

- Processed material: refined material, metal, plastic, wood

- Preforms: materials whose form, dimensions, and composition have to some extent been standardized, such as profiles of plywood or various composition boards. Preform materials may come in the following forms: elongated, such as tubing rods, wire, and 2x4s; sheet, such as plywood, sheet metal, and screening; modular, such as bricks and precast building parts.

- Parts: miscellaneous parts which have been standardized, such as fasteners, hooks, hardware, etc.

- Components: engineered, mechanical, electronic components, such as light bulbs, motors, switches, relays, etc.

- Object: defined here as any entity which, in conjunction with other entities and environmental aspects, can become an integral part of an operational system, such as, furniture, appliances, and tools. This category includes everything in our classification of the man-made object.

- Operational system: constitutes an interrelated and inter-acting group of objects and environmental aspect which can be operated (used) to effect certain desired changes, such as the various objects and environmental aspects required to type a letter or slice a loaf of bread.

The listing can, of course, be extended; however, the above is sufficient to indicate the range of the matter sector of the transformation process.

Uniformity of Materials

Before considering the forming and design sector of the transformation process, some further information about matter sector needs to be discussed. Industry provides the designer with a great variety of materials (represented in the matter sector). Their composition, preforms, parts, and components are generally standardized. Such standardization serves to obtain economical benefits in the production of these items and also the final product using these materials. However, from the utilization point of view this may not necessarily the best for the product. Standardization means uniformity: compositional and form uniformity. Such uniformity of material used in the designing of a product, may merely approximate the actual requirement demanded by utilization. Uniformity tends to influence the design of a product, of which the designer must be aware.

Thus the size of the tubing used in the frame of a bicycle, for instance, is determined at the point of greatest stress. Because of the uniform cross section of the tubing, excess material is placed at points of minor stress. Similarly, when tubing is used for the legs of a table, the greatest stress will be at the point where it is attached to the table and less at the lower portion of the leg. Moving the excess material towards the point of greatest stress would strengthen the product without using any more material. Of course, there may be other factors involved in the choice of a particular, e.g., preform. However, an awareness of these facts may enable the designer to improve the design and/or conserve material, should the opportunity present itself.

Some time ago a manufacturer followed this line of reasoning in making a new kind of face towel. Towels are generally woven uniformly throughout; every section is of equal strength and thickness. However, when a towel is used, every part does not receive the same amount of wear and tear. The center portion is generally used far more than the rest of the towel. Therefore, if utility is a prime determining factor in the design of the towel, and wear and tear are taken into consideration, a somewhat non-uniform weave—i.e., a reinforced center—would better fulfill the requirements, and doing so without using any more material. Also a manufacturer of tubular furniture, in following this line of thinking, produced tables with tapered tubular legs.

The Forming Sector of the Transformation Process

The forming sector of the transformation process refers to the various means used to transform the materials and entities which are part of the matter sector. These means may be man's bare hands or the extension of those hands—tools, including manually or mechanically powered machines, controlled directly by man or automated and computerized.

In the past, the selection of materials and a manufacturing process for a given product often occurred by default, since they were not specifically selected or developed for that particular product. Thus the furniture industry, for instance, was a wood industry. Selection was consequently not in the hands of the designer but was determined by the industrial and commercial situation in which the designer worked. In recent decades, however, certain changes have taken place. New materials and manufacturing processes are now available; and many products can be produced in different materials or in a combination of materials. Thus the designer also may be faced with the determination of the appropriate material and process for his or her design. This raises the question, what is the basis on which the appropriate means should be selected? Although huge amounts of data and information are available to the designer, so far as specific materials and processes are concerned, we lack a comprehensive survey encompassing all available means—a survey which would make comparison possible. We need a unified approach to the vast field of material conversion, a pattern encompassing all known and perhaps some as yet unknown manufacturing processes. Such a survey would allow the designer to select the most appropriate material and manufacturing process for a given design project.

The Design Sector of the Transformation Process

The design sector of the transformation process is the primary sector, controlling every phase of the process and thus the final result.

The three faculties of the designer involved in a design procedure are:

- Skill: the sufficient ability and dexterity to manipulate the matter involved in a design procedure. These must be possessed by the designer or otherwise be at his or her disposal.

• Knowledge: both the technical knowledge regarding the physical property of the matter being transformed or utilized and the theoretical knowledge regarding the man-made object

• Conceptual: the personal ideology, attitude, motivation, inclination, or *Weltanschauung*

Any task undertaken in transforming environmental aspects involves these three faculties: adequate skill, pertinent knowledge, and some notion of how to proceed with the task. Even such a simple task as sweeping a floor demands the ability to manipulate the broom, an understanding of the broom's action and its resulting effect, and some notion of how to proceed sweeping the floor. Each person approaches the task of sweeping the floor in a somewhat different manner.

Although these three faculties are very much interwoven, it is the conceptual faculty which actually controls the whole procedure. It not only determines what skills and kinds of knowledge a designer has, but also to what extent they are being applied in a given design procedure. We will consider each of these faculties further.

The Skill Sector of the Design Process

The skill sector of the design process refers to two kinds of skills:

• Graphic skill: the ability to sketch, draw, and render. This includes freehand drawing, mechanical drawing, perspective, and rendering in various media.

• Technical skill: the ability to work various materials in order to make sketch models, tryouts, mock-ups, and replicas

As stated before, the designer must personally possess these skills or must have someone perform them under his direction.

There seems to be general agreement concerning the kinds of graphic skill essential for the designer. It will not be necessary to refer to these specifically, except, referring to teaching of freehand sketching. The inability of so many students to draw freehand sketches is a problem which is often discussed. Whereas some students (generally only a few) seem to have what might be called a native ability to sketch, the majority have considerable difficulty with this aspect of designing. The claim that the latter group lacks talent as designers seems a facile explanation. One has to consider the possibility that a particular approach and method used in teaching freehand drawing may be inappropriate for certain students. Schools have tried to overcome the difficulty by prolonged sketching exercises intended to improve motor control. However, such exercises do little to teach the student how to perceive visual images correctly. An analysis of what takes place when a visual image is perceived may shed some light on this problem and may suggest a new approach.

The student is not a camera but someone who experiences visual images as an integral part of daily living. There is a difference in this daily perception of images and the way we perceive an image we are trying to draw. The following demonstration will illustrate this difference and the way it may affect a student's ability to sketch.

The instructor holds a rectangular piece of cardboard in front of the students in a vertical position. The students are assigned the task of drawing the shape of the cardboard. This is followed by a second assignment in which the instructor uses the same piece of cardboard but holds it at one corner in such a position that the opposite corner slants towards the students. The visual image is now either a rhomboid or a trapezoid, depending upon the relative position of each student to the cardboard. Again, the students have to draw what they see. Most likely some will have difficulty with this assignment. But why? To sketch the cardboard in a vertical position the student only has to consider the proportion of the two sides. Since this proportion is known to the instructor, it can readily determine how close the student has come to the correct figures. (At this point, we will ignore the difference in position of each student relative to the cardboard.) No difficulty is experienced in the first assignment. Why, then, does the second position of the cardboard create problems for some students? The difficulty apparently arises from the fact that, in this position, the visual image is no longer congruent with the actual physical shape, as it was in the first position. This discrepancy seems to result in a conflict. Although the student does not "see" the true physical shape, it is nevertheless apparently experienced, regardless of its position. A similar tendency can often be observed in the drawings of children. At certain stages in their development they may draw a table, for example, not as it visually appears but closer to its true physical characteristics. Thus the top might have a rectangular shape, although it could only be seen this way from a position above the table. Or the table might have four legs sprawling out from each corner, despite the fact that the beholder can never see them in this position. This seems to indicate that a child draws not what it sees but what it actually experiences. This ability to "see" the real rather than the visual image is not necessarily a negative trait for someone who has to develop three-dimensional objects.

A further study of the discrepancy between the physically real and the visually real might serve to shed more light on this problem and lead to a new approach in the teaching of sketching. The difficulty encountered by students seems to be due to their three-dimensional spatial orientation as opposed to the two-dimensional orientation of a drawing. This suggests that familiarity with the rules of perspective prior to sketching exercises might enhance their ability. It would enable them to grasp intellectually the change of a visual image relative to the position of the observer.

Technical Training for the Designer

Whereas graphic skill is commonly considered to be essential for the industrial designer, technical skill is a different matter. In fact, it sometimes has been maintained that manual training is unnecessary, because model makers take care of it in the field.

Today, most schools offering an industrial design major maintain some shop facilities equipped to handle a variety of materials and forming processes. Their prime purpose is to make try-outs, mockups, and possible prototypes of a given design and, only incidentally, to train students. The student who works in these shops often has to rely on his or her own resources, since little or no specific instruction is offered. Such a limited approach gives the student only a spotty familiarity with materials and processes. On the other hand, the craft-oriented training offered at the Bauhaus also is not the proper training for today's industrial designer. Not only were students there limited essentially to one material, but their orientation was still primarily towards handicraft. This kind of training can now be left to the material-oriented craft schools that train the studio craftsman.

Directly related to and perhaps more important than manual training is the student's technical knowledge of various manufacturing processes. It has been said that the designer is the liaison between the manufacturer and the consumer. He is also the liaison between the forms of his design and available manufacturing processes.

Every manufacturing process leaves an imprint upon a product. Some imprints may be acceptable from a utilitarian as well as a visual point of view; others may not. Certain processes have their typical forms—like those, for instance, of wood and metal turning. The designer who is concerned about a product's utility may question the extent to which the forms resulting from certain manufacturing processes meet the actual demands of utilization.

If neither the craft-oriented Bauhaus approach nor a merely incidental contact with materials and processes can be considered sufficient for today's designer, what would be a proper educational approach to the technical aspect of product design?

The curriculum in some schools includes technical courses that cover various industrial processes. It seems to me, however, that the approach is from a purely technical point of view not directly related to the actual design process. To overcome this limitation in my own teaching, I introduced the Application Course Series, which relates various manufacturing processes directly to design procedure. More will be said about this design series in Chapter 7.

While it is impossible, indeed unessential, to thoroughly familiarize the student with all materials and manufacturing processes, it is possible to give the vast field of material conversion a pattern. The establishment of such a pattern would enable the designer to oversee the entire field, to choose appropriate materials and processes for his or her design, and to be adequately prepared to cooperate with the engineer. Although some work has been done to establish a comprehensive pattern of all material conversion processes, the result is still insufficient. It remains a task for the future. In the meantime, as an initial step in the right direction, we can establish an educational program that covers all major material conversion processes, from simple handicraft to hand-operated, automatic, and computer-controlled machines. I followed such a program in my teaching and—this is the important point—dealt with the various materials and processes not from a purely technical, but essentially from a design point of view. More will be said on this subject in Chapter 7.

The Knowledge Sector of the Design Process

The knowledge sector of the design process refers to two kinds of knowledge: knowledge of the materials and manufacturing processes and knowledge regarding the various aspects involved in the utilization of products. The former has been dealt with in the preceding discussion on technical training. The latter aspect is usually dealt with by offering a variety of courses in science, history, art, etc. Although these courses are important, serving the student's general education, they must be supplemented by a comprehensive study of the man-made object and environment, a study which is directly applicable to the designing of products.

5

The Conceptual Sector of the Design Process

The most important aspect of the design sector is the conceptual faculty of the designer. Although no task can be accomplished without the involvement of the other faculties, it is the conceptual faculty of the person directing the procedure that is paramount. The designer's ideological makeup, concepts, and values initiate and sustain the process of designing: they determine approach and direction and, therefore, the final outcome. It is the conceptual faculty which makes a designer do what he does and determines the precise manner in which he does it. The conceptual faculty, in short, determines both ends and means. Appropriate skills and technical knowledge are indispensable in any task. What kinds of skills and knowledge a designer acquires and applies will depend on the designer's inner makeup—and the task itself, of course, is radically conditioned by the way in which it is conceived. However superb, mediocre, or insignificant a design may be, it reflects the designer's inner nature and whether or not he or she is aware of a personal *Weltanschauung* (personal philosophy of one's own world). It plays a major role in any design task conceived and undertaken.

In design schools that preceded the Bauhaus, the conceptual aspect of a design procedure was considered to be a matter of personality; consequently, it varied according to the instructor. The student could often choose the instructor best suited to his or her way of thinking. Considerable changes occurred in the twenties with the establishment of the Bauhaus. It initiated a new approach in dealing more specifically with the conceptual aspects of design education, as indicated by the following statements: "The Bauhaus is consciously formulating a new coordination of the means of construction and expression," and "The Bauhaus strives to coordinate all creative effort."[31] The intention of the Bauhaus introductory course was not to impart certain skills or convey knowledge but to affect the conceptual faculty of the student—to arouse creative potential. Today, many schools of design have adopted some kind of introductory course similar to, or a variation of, the original course at the Bauhaus. Some vary only in the nature of their assignments (see the discussion on John Arnold and Albert Szabo in Chapter 6). All, however, seek to stimulate the student's creative efforts.

Some Common Misconceptions in the Field of Design

Industrial design is generally considered to belong to the visual arts together with architecture, interior design, commercial art, painting, and sculpture, thus giving the impression that the prime emphasis when a product is developed is on its visual aspect. This generalization of a number of diverse activities may be the result of design education having originated in schools of art. The difference between painting or sculpture and industrial design is obvious. Paintings and sculpture are essentially experienced only through confrontation; i.e., someone directly facing the work in order to contemplate and appreciate it. This action takes place only at certain times. In contrast, a product, room, or building is experienced over a longer period of time, since it is an integral part of our life. We do not sit and stare at a product or an environmental detail as we would at a painting or sculpture. Although a person may at times concentrate on a specific product or environmental aspect, in general this is done only in order to evaluate it or consider its possible use. It is the task of the designer, first of all, to create a satisfactory physical experience for the user. Although the visual aspect should not be ignored, it is secondary to the appropriate solution of the physical aspect. In view of this, it seems odd to refer to industrial design as a visual art. It would be more appropriate to consider it a separate field between art and engineering.

The "Fundamentals" of Design

The notion of design fundamentals is closely related to the idea that product design belongs to the visual arts. Gropius referred to the "means of individual expression" *(Ausdrucksmittel)* and to rhythm as the *"urspruengliches Ausdrucksmittel"* (primary expression), as well as to light and dark, color, matter, tone, proportion, space, and their interrelationship. [32]

In recent times, line, plane, volume, texture, and color are often referred to as design fundamentals. It is interesting that these "fundamentals" are visual and seem to emphasize the visual aspect of product design. But, although these are in one form or another part of a product, they cannot be considered as fundamentals on which the designing of products is based.

The Notion of Self-Expression

Self-expression refers to a deliberate effort by a designer to express his or her personality in the designing of a product by using forms or details which happen to be intriguing. This notion is perhaps based on the belief that the product is the place where the "artist" in the designer comes into play, where his or her "signature" appears. It is a superficial notion and most likely works against the quality of the product.

The designer actually expresses him or herself in any design process but merely as an incidental by-product; not in some mannered application of arbitrary forms but by remaining true to the demands of the given situation and to the approach used in solving the design problem.

Other Common Conceptions

In 1963 R. H. McKim introduced the term *visual clarity* for a number of different aspects of forms.[33] Although the term is new, the concept for which it stands is not. In one form or another, this concept has been familiar to me since the beginning of my design education. McKim listed four types of visual clarities: visual clarity of function, of structure, of the producing tool, and of material.

Visual clarity of function means that the form of a product or building expresses its function. At first there appears to be some logic in the notion that an object should express what it actually is. Does it not make sense, for example, that a telephone readily reveals what it is, that a car represents swift movement and a chair comfort, and that a building is clearly a bank, a church, or whatever else it may be? If this is the intention of the designer, the question arises, what are the consequences with respect to the form the designer has to use? The fact is, in order to design an object that expresses its function; the designer has no other choice but to use forms which are familiar to the beholder. He would have to superimpose forms of earlier telephones, for example, on a new telephone, even if the new components demanded a different form. Obviously, such an application of prior forms not only would be superficial but would impede further advancement.

Another question arises, why should an object communicate its function? If this were possible it would perhaps be useful in an initial contact with an entirely new product, but any further communication thereafter would be redundant. Since such an initial confrontation does not take place in complete isolation but within the operational system and prevailing mental climate of which the user is a part, it would be superfluous for an object continuously to communicate its function.

Of course, some objects require the use of so-called self-suggesting features in order to provide operational directions to the user. These features are not really "self-suggesting" but are merely carryovers from earlier objects familiar to the user. I am referring essentially to activating devices, such as turning knobs and press buttons. Such features are generally transmitted in successive generations of product development. It can be said in a general sense that an object cannot be designed in such a way that it expresses or communicates its function because it lacks the prerequisites to be a communicative device. The fact that we generally recognize an object is due merely to a previous direct or indirect association with it.

Visual clarity of structure suggests that an object should reveal its structure—its inner build-up— for aesthetic reasons and/or to allow us to experience its structural soundness. With respect to the latter, what has been said about visual clarity of function also applies here. An object cannot be formed in such a manner that it suggests structural soundness. If a beholder seems to experience such soundness, it is only because of a previous experience with similar structure forms. To design an object or building which not only is, but looks structurally sound would have serious implications for the design itself. It would impede the development of a new product and the proper use of new materials and processes. This brings to mind a discussion regarding certain features of the Bauhaus building in Dessau designed by Walter Gropius. The balconies of the dormitory consisted of relatively thin concrete slabs, about four to five inches thick, which projected approximately four to five feet from the wall. Only one light tubular railing afforded protection. At the time, it was argued that these balconies did not look safe, although they had obviously proven their safety. To design balconies that both looked safe and were structurally sound, Gropius would have had to make use of structural features which were familiar to the beholder; i.e., which had been used in the past. Obviously this would have

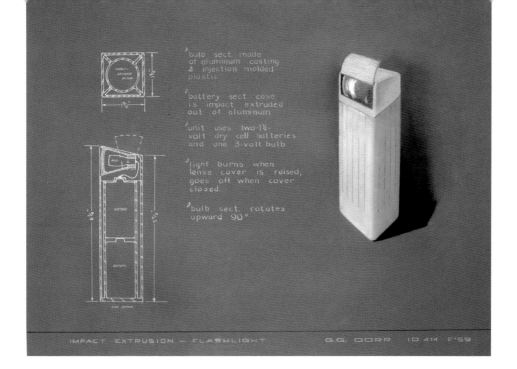

deterred development. In order to satisfy the need to feel safe, one must, when confronted with a new structure, experience it as safe. The structure itself cannot convey this to the beholder.

The reference to structure as an aesthetic experience is essentially a new idea and applies to recent structures which use a minimum of materials, allowing visual penetration and the experience of the "inner working" of a unit. Whatever that experience might be, it affects the beholder merely in an incidental manner. There are some structures which dominate a given unit: the form is merely the result of that particular structure—for instance, the mast that carries an electric high-tension wire. Nevertheless, in the case of most of our products and buildings, it is the form which affects us directly; structure serves to support the form.

Visual clarity of the tool suggests that an imprint on the form of an object caused through the use of certain tools and processes can contribute to its aesthetic quality. It is a fact that no tool or process can be used without leaving some mark or imprint upon the form being produced. Certain manufacturing processes that produce forms which are typical of these processes come to mind—for instance, the potter's wheel and the spinning lathe. Because they produce characteristic forms, they only can be used for specific kinds of objects that have a restricted range of application. There are other processes which have a wider range—for instance, stamping and extrusion—while still others—such as casting and injection-molding—have an even broader range of application. Whatever the range of available forming processes, the general trend in the development of products seems to be that a process which does not satisfy the requirements for utilization will sooner or later be modified or replaced by another process. In other words, there is a constant tendency to modify and develop new manufacturing processes which will adapt the object better to the requirements of utilization (fig. 13). This trend

has resulted in objects which, apparently, do not show the marks of their production process.
The ultimate use of a product is the cardinal factor which determines the form of that product.
To design means to facilitate the utilization process by exploiting the available means
of production.

In the past and occasionally still today, objects were sometimes designed to use forms deliberately evolved from certain manufacturing processes. Some of the earlier Bauhaus products are
examples of this. For instance, Marcel Breuer created a chair in which all the members have the
same cross-section and all the joints are at right angles (fig. 14). Both the uniform cross-section

of the members and the right angle joints serve solely to simplify a particular manufacturing process that contributes little to the utility of the unit. Another example is a teapot designed by Theodor Bogler, which was described as a typical object intended for mass-production (fig. 15). These designs, of course, have to be viewed within the context of their time—a time when designers were striving to overcome the prevailing handicrafts mentality and just beginning to accept the machine as a means of production. It is therefore understandable that in the initial phase of this development there was an overemphasis of the machine, which was soon overcome. In view of this, it seems rather strange to refer to these and other early Bauhaus products as "functional designs" (as is often done), even though certain forming processes largely determined not their function but their form.

Visual clarity of material refers to the prevailing notion that materials should be used according to their "nature." As I have already indicated, this raises a number of questions: What is the nature of a given material? Is there a correct and incorrect way to use materials? And is it possible to employ any material in a manner contrary to its physical nature when you design, produce, and use a product? This aspect has been dealt with in Chapter 4.

Developing New Design Concepts

The design concepts referred to in the preceding section were developed in part without deliberate effort by designers who were active in the field. Now, however, we seem to be entering a phase in the development of the design profession in which more and more designers are concerning themselves with the conceptual aspects of design activities.

The development of new design concepts that fit present times is obviously a difficult task, for the source of one's personal ideology lies deep in the physical and social milieu in which one lives, in the activities one has undertaken, and the experiences one has had. To further develop such a deeply rooted, almost inherent ideology through formal education is obviously not easy.

Today's educational approach to this aspect of design activities is essentially inadequate for the problem. As a result, this all-important facet of design activity, which directly or indirectly controls the quality of any design, largely develops independent of formal design education.

What are we currently doing to develop the conceptual faculty of our students? Our efforts to inculcate skills and technical knowledge are readily apparent in any educational design program. One can point to a number of specific courses in these areas and justify them in terms of their value to the practice of design. But courses which seriously and thoughtfully seek to refine and enrich a student's ideological development—his broad conceptual faculties and personality in general—are not so easy to find. To be sure, most design schools have introduced miscellaneous courses in the humanities, supplemented perhaps by a variety of lectures by visiting designers, artists, scientists, historians, and the like, but this agglomeration of unrelated courses and lectures is but a half-step in the right direction. What is offered in such fragmentary fashion can be dealt with by students only as one of many factors impinging on their personality—on their conceptual system. These fragments may complement the student's personal makeup, but they may equally well contradict it, thus rendering what is offered ineffective or even disruptive. In any case, whatever the effects, they can only be random and incidental. Although this fragmented approach is perhaps the best we are capable of at present, we should realize that it taxes the students with far more than they can be expected to cope with. This approach to education assumes that the student will be able to do what educators themselves have not been able to do. If an educational institution cannot discover the broad ideological or conceptual bases which would inform, relate, and vivify the miscellaneous facts we dispense, how can we expect our students to absorb and organize such an assemblage of information and unrelated ideas and transform them into a consistent and powerful personal ideology?

Failures in design are due neither to a lack of manipulative ability nor to a lack of technical knowledge but rather to our inability to comprehend as an organized whole the great variety of spatial and temporal aspects involved in a design situation, particularly in the design of mass-produced products. Personal "good intentions" are of little avail in such complex situations; what is required is the ability to comprehend these complexities. Although man is naturally limited in his ability to grasp complexities, he is remarkably able to devise ways and means of supplementing his natural ability by extending his grasp. Such means consist of the conceptual organization and patterning of the "territory" of concern—making a kind of "map" or conceptual model depicting the nature and interrelations of the environmental aspects involved.

Consider the following simplified analogy: a man arrives in a city that is entirely unfamiliar to him and whose natives speak a language he cannot understand. Even if this man should possess an unusually acute sense of direction, getting from one place to another in the city would involve him in a crude procedure of trial and error. Even after a good deal of effort, he would still only be able to move efficiently within the territory with which he had made himself familiar. Any deviation from familiar routes would require a new process of trial and error because he could not not have learned much from his previous experiences that would be useful in the new situation created by moving beyond the familiar. Unless this man were able, as he engaged in one search after another, to form some notion of the whole that he could then use to guide further exploration; he would either have to restrict his wants to those which could be satisfied within the limited area he had come to know, or he would have to resign himself to the continued inefficiency of his trial and error method. The whole situation changes drastically, however, as soon as we provide our wanderer with a map: even on the first attempt he can now move with relative ease through the entire city. The map enables him to comprehend the complexity of the city at a glance and the interrelationships of its details. Consequently, such a map, not only gives personal satisfaction and helps him to meet his immediate needs, but also greatly increases both the frequency and the variety of the experiments in travel he is able to undertake and assimilate, thus resulting in expanded horizons and the general growth in well-being that accompanies such an expansion. Many points of interest to design education emerge from this analogy. Let us consider just the following: the map was developed independently from the particular problem encountered by the man in finding his way; it is transferable and could be used by many individuals to solve their unique travel problems; yet "public" and general though it may be, the map supplements each individual's natural abilities, particular interests, and personal inclination. Finally, although it is systematic and analytical, the map ultimately extends the individual's "creative" ability—that part of mental activity that it has all too often been asserted conceptual learning would stunt or restrict.

This analogy is an over-simplification when compared to the complexity of the man-made environment with which the designer has to cope; yet it suggests the value of holistic conceptual systems as devices which can extend our limited natural abilities. Actually, the vastly greater complexity of the man-made environment in which the designer finds himself trying, as it were, to move about, only can serve to emphasize the urgency of the need to provide him with some means of assistance.

The Bilateral Approach

Because form is recognized as an important attribute of any product or entity, a designer is often referred to as a form-giver. But form—the outer boundary of a product—is not the only concern of the designer. As has been said before, any product or entity possesses three basic attributes: form, structure, and position. The task of the designer developing a product is to determine the specific characteristics of each of these attributes.

This approach applies primarily to enclosure-type products, such as appliances and electric devices, but also to buildings. Although appliances and buildings differ greatly in size, function, and complexity, they have one thing in common: an inner core (operational structure), consisting of an interrelated system enclosed by a form and positioned in a given environment.

Let us illustrate the interrelationship of structure to form to position of enclosure-type products and buildings by considering their operational structures. The mechanics of a watch, the inner working parts of a telephone or a toaster, and the various spatial arrangements of the rooms, halls, stairways, etc. of a building all represent the operational structure that makes a unit workable. The initial task in designing an operational structure is to obtain a workable unit capable of producing a certain specified result. This applies not only to the engineered components of a product but also to the inner arrangements of a building. The development of such inner structures can be illustrated, for example, by the electric bulb and the telephone. The initial goal of Thomas Edison in developing the incandescent electric bulb was to obtain a unit that would emit light. Alexander Graham Bell's intention was to construct a unit that could transmit a voice over some distance. Once these initial goals had been met with a certain degree of satisfaction, further improvements of the units began. We will not deal here with the successive stages that took place in improving these components. Instead we will concern ourselves with the general improvements of the relationship of the component to the user and the readily observable trend toward greater compactness of the engineered components. These two aspects strongly influence the configuration of an enclosing form; a good example is the reduction in size of electric bulbs, motors, and other components since their first introduction.

In the development of products, it is generally the task of the engineer to develop the inner components and of the designer to determine the outer form. In architecture, one and the same person generally develops both the inner spatial arrangement and the outer form. Certain engineered components of products actually can be used in the form developed by the engineer. Thus an electric bulb can be used without a fixture, and a telephone can be operated without its enclosing shell. Nevertheless, these components have to be considered unfinished products requiring further refinement. This is essentially the task of the industrial designer who relates a unit more appropriately to the user and his environment. In the total development of a given product, from its initial inception to its final place of utilization, the industrial designer comes last in the line of successive phases involving a number of professions.

The evolution of a product may be said to begin with the location of raw material by the geologist, biologist, or technologist and proceeds through the various phases of processing that material, manufacturing the product, and finally incorporating it into the environment. In this intricate development, each profession makes its own contribution. The designer has a unique position in that, being last in line, he becomes the liaison between industry and the consumer. His task is not only to be familiar with the long line of successive stages but, since he is more intimately aware of the utilization process, to feed back to industry all new requirements that become apparent as the product is put to use.

We can further characterize the difference between the role of the engineer and that of the designer. The engineer seeks to achieve what might be referred to as "laboratory efficiency"

fig. 16 (below)
Samson Toaster
fig. 17 (opp. page)
Gruen Curvex Wrist Watch

(a measurable quality of performance), whereas the designer strives to obtain high overall performance of the product in its final place of utilization. The latter is generally an immeasurable quality. At times, the designer may have to reduce the laboratory efficiency of a component in order to obtain high overall performance. Thus the rated laboratory efficiency of an electric bulb (the lumen/watt ratio) refers to the naked bulb. Adapting the light source of a bulb to man and his environment by means of a fixture may reduce its overall laboratory efficiency but improve its overall performance in its place of utilization.

How does the designer proceed in attempting to achieve such overall performance? Obviously he must begin by familiarizing himself with the nature of the component, its operational system, and the arrangement of its details. These details are generally arranged so as to obtain maximum compactness in order to save material, labor, and space. This is essentially prompted

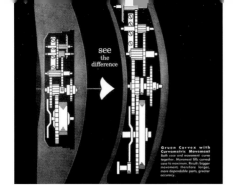

Gruen Curvex with Curvometric Movement Both case and movement curve together. Movement fills curved case to maximum. Result: bigger movement; therefore larger, more dependable parts, greater accuracy.

by production and distribution considerations rather than the requirements of utilization. But the strongly consumer-oriented industrial designer, concerned with the performance of the product when in use, may consider modifying the arrangement of an engineered component to adapt it more appropriately to the user and his environment. His goal is to make the product serve the consumer, and all effort spent in designing and producing is directed towards that goal. Generally speaking, most engineered components allow considerable freedom for such a rearrangement without interference with the function of the operational system.

The factors determining such rearrangements are referred to here as the outer constants of the product. I remind the reader that the term outer constants refers to those aspects of a design procedure that have to be accepted as they are and are commonly referred to as the parameter of the unit being developed.

There are basically three types of outer constants:

• Man: the person or persons operating and/or coming in contact with the unit.

• Other objects: environmental aspects which have an operational, physical, and/or visual relationship to the unit.

• Prevailing conditions: the various conditions which, in one way or another, affect the operation and/or physical integrity of the unit.

We will not deal with the issue of the outer constants in detail here but merely illustrate their possible influence on both the specific arrangement of the details of the components and its enclosing form. See figures 16 and 17. In figure 16, the Samson Tandem Toaster shows how the inner arrangement of the components resulted from consideration of the elongated space that might be available on a counter; and outer constants. In figure 17, the Gruen Curvex Wrist Watch shows how the specific arrangement of the mechanical component is influenced by the curvature of the arm; and outer constants.

These samples illustrate types of rearrangements rather than mere "compactness." They also indicate that it is advantageous for the designer to cooperate with the engineer, whenever possible, in the final development of the engineered component, particularly in the arrangement of details.

fig. 18 (below)
United States Capital,
Architect of the Capital
fig. 19-20 (opp. page)
Typewriter, W. E. Lerdon,
1958

We will now consider the factors which determine the enclosing form of a product. Whatever form a designer creates; it is essentially the result of a personal ideology. This is true regardless of whether the designer has developed his or her own concepts or adopted them. Although ideologies vary from designer to designer, there are certain concepts that are generally accepted. We shall illustrate the influence of these concepts upon the development of form by considering buildings as well as products and by using the Bauhaus as an historical point of reference.

Many buildings of the pre-Bauhaus period were designed with a strong emphasis on the facade and outer form. The underlying concept of this approach, when carried to an extreme, was to superimpose upon an inner core (operational structure) an outer form conceived relatively independently of the characteristics of the inner core. Thus the often-used symmetrical outer form revealed little or none of the internal arrangements. Sometimes windows, for example, were placed without respect to the internal requirements, or so-called blind-windows were used merely to obtain a symmetrical design on the façade. This approach can be referred to as an "outside-in" approach, because the designer appeared to proceed from the outside to the inside of the building. This does not imply that the interior was ignored but that the outer form, especially the

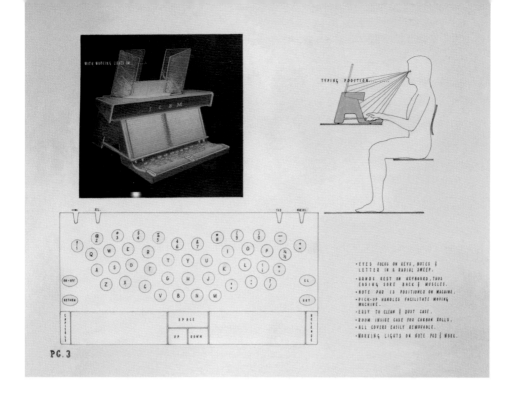

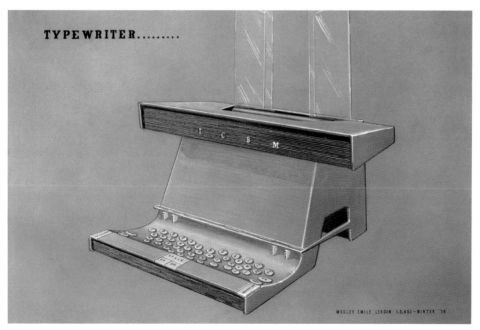

façade, dominated the design, not only with respect to the inner core but also to the environment in which it was located (fig.18).

In the early period of industrialization, a few products also reveal a tendency, although a minor one, toward an outside-in approach (fig. 19 - 20). Later, with the advent of the radio, many, if not most, receivers were designed with an outside-in approach (fig. 21). These outer forms, like those of the buildings mentioned above, were generally symmetrical and often dominated the environment in which they were placed.

As far as products are concerned, however, this outside-in approach was a passing phase. Products soon began to reveal more of their inner cores. The industrial designer, as we know him today, did not exist at the time; it was the engineer who developed both the inner component and the enclosing form. The sole aim in choosing an enclosing form was to protect the component and to prevent foreign matter from interfering with the operation of the unit. The inner component dominated the outer form, which essentially followed the general contour of the component.

Do you think the Volkswagen is homely?

The Volkswagen was designed from the inside out.

Every line is a result of function. The snub nose cuts down wind resistance. The body lines hug the interior workings. Nothing protrudes.

One Briton called the Volkswagen "a marvelous economy of design."

An American owner put it differently. "It's funny," he said, "how she grows on you. At first you think she's the homeliest thing you ever saw. But pretty soon you get to love her shape. And after awhile, no other car looks right."

The VW defies obsolescence. You can hardly tell the doughty shape of a 1950 model from a 61. To suggest altering it is heresy to owners. (Would you change the perfect form of an egg?)

But we are continually making changes you cannot see. Example: a new anti-sway bar eliminates sway on curves. Over a hundred such changes since 1950, but never in the basic design.

Is the Volkswagen homely? It depends on how you look at it (and how long).

The design approach employed in these industrial products differed considerably from the approach in architecture. Whereas the latter was dominated by the outside-in approach, industrial products were already being developed by the inside-out approach. Although these products were multiplying rapidly, they appeared to have little direct influence on the form of buildings. It was only with the advent of the Bauhaus, which followed the inside-out approach, that a switch finally occurred in the approach used in architecture. The outer form of buildings now began to evolve from their internal structures. The former outside-in approach was regarded as superficial and derogatorily referred to as "façade architecture," and the newly proclaimed inside-out approach was lauded as the logical one.

For—as it was argued at the Bauhaus—was not the inner core of the product or the interior space of the building the sole reason for developing the unit? Should, therefore, this inner core also determine the outer form? The inside-out design became known as "honest," because it did not hide the inner parts. Products and buildings began to expose internal aspects, as Walter Gropius' 1914 design for the Werkbund building in Cologne, Germany, illustrates so well (fig. 22). At the time, the inside-out approach was widely practiced both in architecture and product design. Indeed, at times it was the criterion of quality in design. Witness a Volkswagen advertisement that states that the car is "designed from the inside-out" (fig. 23). Or examine Le Corbusier's statement that "it [architecture] must grow from the inside out, the concepts must be biological, not static. A beautiful seashell is not a façade; it is a shell—this is the essence of architecture." [34]

References to natural objects are often made in justification of the inside-out approach. Nature may indeed provide us with valid criterion, but the important point is to interpret nature correctly. The statements such as the one made by Le Corbusier assume that a natural entity evolves only from the inside out. Is this not a somewhat superficial observation of nature? Does nature evolve the form of its products solely from within? Is the shape of the seashell referred to by Le Corbusier exclusively the outcome of its inner biological structure? The fact is that the configuration of the outer form confining the inner structure (in this case, the animal) actually results from an interaction between the inner nucleus and its external environment. It can basically be said that such external form only can be the result of an interaction of inner and outer forces. A form cannot exclusively evolve from the unilateral action of what it has inside.

Entities always exist in an environment; and this environment impinges upon and affects the form, even if the environment should be a vacuum in the universe. Hence, a living entity is part of the biological ecological environment, just as an inanimate entity is part of the physical ecological environment. Whether the entity is biological or inanimate, it owes its integrity to an interplay of offensive and defensive forces of action and reaction. Although an entity might be biologically "dead" (for instance, a pebble on the beach, a car in the street, or a planet in the universe), it is never physically inert or inactive. The inside-out concept applied to an entity leads to the absurd conclusion that it has an "absolute integrity."

What, then, accounts for the great popularity of the inside-out approach? Like the outside-in approach, it is "object-oriented." Essentially it focuses on the product to be developed and relegates related factors to a secondary position. However, the simple fact is that any product is useful only in conjunction with other objects and environmental aspects. An object per se is merely a detail of an operational system, which in turn is part of a larger system. This being the case, we must move away from "object-orientation" towards a "system-orientation." In Chapter 2, we dealt with the significance of such a reorientation. It is worthwhile to note that, for the designer, it would have a profound impact on the outer form of products and buildings. The external factors of a unit being designed, which were previously treated in an incidental manner (as parameters), would now become an integral part of the design procedure.

Before considering the effects of the external factors on the outer form of objects and buildings, let us turn once more to the component, the inner core, of products. The component of a product that is being enclosed has certain dimensions and configurations resulting in a certain "volume-form" (i.e., the general form of the component irrespective of the details). Quite often this volume-form, in accordance with the inside-out approach, determines to a large extent the dimensions and particular form of the enclosing shell. However, this does not necessarily justify the form. The reason for choosing a particular form may well be merely the relatively large size of the component. The enclosing shell does not necessarily have to conform to the dimensions and contour of the volume-form. This point will be considered further.

In the general development of engineered components, one can observe a trend towards ever smaller and compacter units. This trend has accelerated since the advent of space exploration, in which the miniaturization of components is vital. This miniaturization in turn affected the components of commercial products. It is interesting to speculate about what will happen to the enclosing shell of our commercial products if the components become smaller and smaller. Will these miniature components still determine the dimensions and configuration of their enclosing form? This could hardly be the case. Let us suppose, for instance, that it were possible to reduce the component of a telephone to the size of a matchbox or even smaller. No doubt, the enclosing form would not be equally small. So diminutive an object, when set on a desk, could be easily misplaced and, indeed, would be difficult to manipulate. The important point is that, should this development of miniaturization spread further, it would undoubtedly mean that the internal component could no longer determine the dimensions and configuration of the enclosing shell. Having thus lost the "vital" factor that previously largely determined the outer form, what new factors could then be found to assist the designer? Obviously, these new factors could only be external to the unit.

Although a shell is developed because a component needs to be enclosed, it must do more than merely cover it. An engineered component is developed in order to bring about certain desired environmental changes. Once the inner component has been obtained, the unit as such becomes a quasi-independent environmental unit, to some extent irrespective of its primary initial purpose. As a mere entity in our environment with which we have to reckon, the unit poses a problem. The succession of problems in a design procedure—i.e., the solution of one problem, which in turn gives rise to another—can be illustrated by the following example. Consider the spanning of a river by a bridge. To reduce the span, two piers are placed in the river. In order to support the load of the bridge, these piers require cross-sections whose structure-form depends on the type of construction and material being used. Although there are various kinds of structure-forms that could support the load, there well may be other factors that have to be considered beyond the load-carrying capacity. For instance, should the river be covered with ice-floes in winter, the opening between the piers would have to be bell-mouthed in order to prevent the ice from accumulating in front of them; the bridge requires a support, and this support in turn makes its own demands. Thus, determining the final form of a product solely by its function and use cannot be justified. In addition, since the object exists twenty-four hours a day, although it may be used only a fraction of that time, it has a visual and physical relationship to man when it is not being used. The question thus arises, what is the form of an object when it is not in use; i.e., when it is dormant? What factors determine the form of products and buildings beyond operational or functional requirements?

Each enclosure type of object, whether a product or a building, has a twofold relationship: (a) to that which it contains and (b) to that which contains it. Hence, this type of object is at one and the same time both a container and a containee [sic]. This twofold relationship thus not only represents the major factor in determining the form of enclosure-type products but is inherent in biological forms that evolve in nature, contrary to Le Corbusier's statement concerning the unilateral development of a seashell. When applied to design procedures, this approach is referred to as "bilateral," because it combines both the inside-out and the outside-in approach. Although the bilateral approach is basically acceptable, it obviously does not answer all the questions that arise with respect to the final outer form of our products. This problem will be dealt with further in a later chapter.

Two Basic Approaches to Design

In recent times, both here and abroad, noticeable efforts have been made toward the development of more systematic design procedures. Such new concepts as "product research" and the establishment of "parameters" or "determinants" are signs of this development. These efforts seem to concentrate directly on the procedure itself. Although this can be considered as an initial step, it is nevertheless too limited. In order to advance our understanding of the design process, it is essential to consider it merely as one of many aspects of our complex, evolving man-made environment. A comprehensive study of the man-made object as suggested in this writing will reveal a clearer picture of the design process. Put briefly, objects can basically be obtained in two ways: they can be found, or they can be made. These possibilities are represented in two approaches, which we will discuss in the order of their historical development: (a) the application approach and (b) the situation approach.

The Application Approach

This approach to design starts with the selection of a certain object or entity and seeks to find an appropriate situation in which it can be useful. It is the use of an object or entity for a purpose for which it was not especially designed. In its purest form—i.e., without any modification— it is the selection ("selection" being a form of design), for instance, of a shell from a beach for use as a cup. In general, however, an object so selected is modified to one degree or another, sometimes only as a rather subtle modification of a detail in a manufactured object intended for a subtle form of application. In general, however, an object so selected is being modified to some existence.

Early man selected certain entities and initially merely changed their positional attribute, leaving the formal and structural attributes as they were found. As time went by, however, these selected forms were modified more and more, until finally the form was completely man made. Many objects, however still retain the original structural attribute—the inner build-up of the form. This attribute remains in general still a matter of selection. We are only beginning to design the specific inner build-up of our products.

Any object or entity can be subject to application. For instance, the use of a chair as a stepping stool or a rock as a seat is a form of application. In the design process, awareness of the possible applications of an object is more than a matter of academic interest; it is of practical importance to the designer. Most of our products include some parts which are merely selected—not from nature but from the vast array of materials and miscellaneous entities available today. The important point is that the application of such preexistent parts may present merely an approximation of the actual requirements of a given situation. In other words, it represents a compromise. In a given design procedure, due to economic considerations the designer may have no choice but to use a certain preexisting material, preform, part, or component etc. The important point is to be aware of any kind of application of whatever nature, in order to check to what extent these satisfy the actual demands of the situation. Should this be a too great a compromise, the designer may consider replacing these selected parts by specifically designed items.

The Range of Potential Applications

The following list merely serves to indicate the range of possible applications that may influence the form of a product:

- Materials: the selection of natural and man-made materials

- Preforms: the selection of standardized materials, such as sheets, profiles, and modules

- Parts: the selection of standard parts, such as fasteners or hardware

- Components: the selection of standard mechanical and electrical items, such as motors, relays, switches, or electric bulbs

- Existing forms: the selection of existing natural or man-made forms, such as a clock in the form of a teapot

- Production forms: the selection of a manufacturing process that leaves a specific mark on the product, such as a potter's wheel or spinning lathe

The Situation Approach

The situation approach is the reverse of the application approach. It starts with the definition of a given situation—an existent environmental problem that requires a solution—and proceeds to the development of the means which can serve in that situation. This approach presented here is essentially for the development of transformer-type objects.

A design procedure of the situation approach can be dealt with in five different ways, i.e., the extent to which the problem is being considered. These are referred to here as "orientations" and are called object-oriented, process-oriented, system-oriented, time-oriented, and ecology-oriented approaches.

The Object-Oriented Approach

Assuming a given situation (design problem), an assignment following an object-oriented approach would refer to a certain object by name. This means, if taken literally, the assignment calls for the redesigning of an existing object, i.e., merely changing form, color, texture, etc. External factors are dealt with only to determine these aspects of the to-be-designed unit.

The Process-Oriented Approach

Assuming a given situation (design problem), an assignment following the process-oriented approach remains essentially object-oriented in that it also refers to a certain object by name. However, besides the form, it seeks to improve its operational procedure and the particular manner, order, and number of steps required in using the object.

The System-Oriented Approach

Assuming a given situation (design problem), an assignment following the system-oriented approach concerns itself, besides the preceding concern, with the whole system of which the particular situation is merely a detail. This is based on the fact that a transformer-type object is being used in order to obtain certain environmental changes. Secondly, it becomes operational (useful) only in conjunction with other objects and environmental aspects. Such an interrelationship constitutes a system involving various objects and environmental aspects.

An assignment does not refer to a certain object by name but rather to the changes which are to be obtained. These changes are specified and the task of the designer is to develop the means that can be used to obtain these changes.

The Time-Oriented Approach

Assuming a given situation (design problem), an assignment following the time-oriented approach considers, besides the usage phase of a given object (commonly referred to as its function), the fact that an object exists twenty-four hours a day. It may be used, dormant, in storage, displayed, repaired, shipped, discarded, or recycled.

Ecology-Orientated Approach

In this assignment the approach itself is part of the assignment. For this the following data are provided.

Assuming a given situation (design problem), an assignment following the ecological-oriented design approach is based on the fact that any given object is an integral part of a rather complex ecological environment. An assignment may call for dealing with a rather specific design problem; nevertheless, the designer must consider all environmental aspects, which, in one way or another, may have some bearing on the particular situation being considered.

Two aspects that determine the specific characteristics of any given object are manufacturing and utilization. Since these two areas are in a constant state of flux, they are being considered not only as they are at a given time, but also how they may be with possible future changes. Changes occurring in one area may affect changes in other areas. To illustrate: the orange-juice squeezer once found in nearly every home has become a rarity due to changes in the citrus industry; and the introduction of pre-pressed wearing apparel may not have eliminated the electric irons, but it has diminished its demand. In addition, the historical development of the kind of object or situation dealt with can be of interest to the designer, as each object has its predecessors as well as its successors.

The above five situation approaches, as well as the application approach, will be considered further in the Chapter 7 on education.

Basic Design Procedure and Graphics

There are perhaps as many different design procedures as there are designers. Nevertheless, there are certain basic steps which every designer has to deal with in one manner or another. These can be referred to as the basic steps of a design procedure, which are often interwoven. Referring specifically to the development of transformer-type products, there are three such phases: 1) the manual presentation of concepts; 2) the development of principles; and 3) the ideation process involved in developing those principles.

Manual Presentation of Concepts

When developing a product, the designer generally makes a great variety of sketches and drawings, each designer doing so in his or her own way. These various sketches and drawings can be placed in five categories: (1) notation; (2) realization; (3) variation; (4) optimization; and (5) visualization.

From a logical point of view, these steps follow each other in the above order; in practice, however, the procedure depends on the working habits of the designer and perhaps also on the kind of the product being designed. Let us consider each step further.

Notation

A notation is a two- or three-dimensional sketch depicting a seemingly feasible conception of a principle that could satisfy the demands delineated in the statement of a problem. It is only a preliminary step toward a viable design. Occasionally, a notational sketch may serve as a means of communication. However, its essential purpose is to solely benefit the designer. Since this is the case, the designer may, if so inclined, ignore conventional rules of drawing, so long as he or she understands what is being done. The sketches may show incorrect proportions, perhaps only because some of the correct dimensions are not known at this stage. In general, this would not alter the essence of the principle and would be corrected at a later stage. What is important at this point is not to waste time indulging in fancy sketches. These should be as brief as possible (perhaps only schematic) but sufficient to depict the essentials of the principle. In working on the subsequent phases, further notation sketches are made.

The purpose of a notational sketch is to enable the designer to grasp and understand a perhaps rather vaguely conceived concept of a principle. Further, to store temporarily the concept, in order to evaluate and to compare it with other equally notated principles. In addition, and this is a rather important point, it can serve to facilitate the ideation process to obtain other principles. This will be dealt with later.

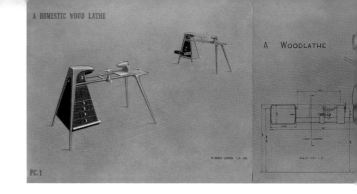

fig. 24 (left to right)
A Domestic Wood Lathe, W. E. Lerdon
A Woodlathe, W. Watkins
A Clock, Unidentified
Outdoor Drinking Fountain, W. E. Lerdon
A Thermometer, Unidentified

The Realization Phase

The next step of this procedure deals with the realization of noted principles. This phase has three aspects: (1) the feasibility of the principle; (2) the consequences of the principle; and (3) the acceptability of the consequences.

Having developed a number of notated principles, the task is now to establish their feasibility; that is to determine for each of the notated principles the approximate dimensions, form, and features required to make them workable. Thus, this step transforms a mere notated principle graphically into a workable piece of hardware. These requirements are the consequences of any particular principle under consideration. This evokes the question, Are the consequences acceptable? Although a given principle may be workable (i.e., able to achieve the desired performance), it may not be satisfactory for a number of reasons. Its dimensions, form, or features may not be appropriate in their relationship to man and other aspects of the environment. Now the task is to determine which of the tentatively acceptable principles deserve further consideration or which should be abandoned? Keep in mind that changes are possible in the following phase.

The graphics of the realization phase are, in general, mechanical-scale drawings depicting the principle as briefly as possible in its correct dimensions and proportions, so far as these can be determined at this point. Whatever kinds of drawings are used, they should depict only the absolute essentials. These are being made solely to enable the designer to understand the specific nature of the principle. Since the realization helps to familiarize the designer with the problem (as well as to establish feasibility and consequences), it is advantageous to make a realization early, perhaps after having made only a few notations. This also aids the ideation process in the development of new principles. It should be stated that a realization is not a mere recording of principles, but a means by which the designer can creatively develop other principles. This will be dealt with in the next chapter.

The Variation Phase

The realization phase results in a number of principles which conceivably would work. However,

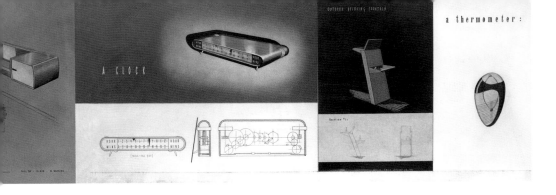

while each principle may well be operative, it may not present the optimum of that specific principle, so far as its physical characteristics are concerned. The variation phase, in keeping the essentials of each principle constant, seeks to establish the optimum solution (dimensions, form, and features) of each principle.

The Optimization Phase

After having developed a number of feasible principles and established their consequences, as well as the optimum of each of their physical characteristics, the task then is to evaluate all obtained principles in order to select the overall optimum solution. This may involve some further development, such as combining certain features and details from different principles into one solution.

Since the sketches and drawings made in the preceding phases are essentially for the benefit of the designer, these graphics can be in the designer's own "handwriting." This is not the case, however, in the following phase, which aids communication.

The Visualization Phase

In the visualization phase, selected concepts are made visual through rendering, drawings, models, etc., in order to present them for further scrutiny, evaluation, and possible acceptance (fig. 24).

The Principles of a Transformer-Type Object

Many design procedures do not pay particular attention to the different principles of this kind of object. They are all too often dealt with in accordance with the personal inclinations of the designer. What are these principles?

We will recall that a transformer-type object is used to effect certain desired changes in other objects and/or in environmental aspects. Objects of this kind involve four basic principles: (1) the operational principle; (2) the mechanical principle; (3) the technical principle; and (4) the relational principle.

These principles represent different aspects of one and the same object. The designer, although aware of all of them, develops each principle separately in the order listed above. When not being considered, the other principles assume a mere supporting role. In the development of a new product, these principles are best dealt with in succession. The operational principle comes first, since it is the actual reason for developing the object. Only after an operational principle has been found can the other principles be developed. Second in line is the mechanical principle, which makes the operational principle possible. This is followed by the development of the technical principle, which is the realization of the preceding principles as an actual product. Finally, the relational principle relates the product appropriately to the environment in which it will be operated.

The importance of being aware of these principles, serves to draw attention to the designers personal make-up. For example, he or she may unwittingly concentrate on one principle to the detriment of the others. Those principles which are not dealt with in a deliberate manner will be determined by prevailing circumstances, according to the immediate situation. A specific product—the rural mailbox—will serve to illustrate these four principles.

The Operational Principle

From a system-oriented approach, the mailbox interconnects the postal system with that of a home system. The operational principle refers to the placement, holding, protection, and removal of mail. Different means can be developed for each of these aspects. Mail can be placed from the front, from above, from below, etc., and can be held upright, flat, on end, etc. The different ways of protecting and removing the mail are likewise operational principles.

The graphic means of presenting these concepts most likely already include certain mechanical and technical aspects. It should be kept in mind, however, that they merely serve to make these concepts visual and understandable, since they are specifically dealt with in following phases. Once a number of operational principles have been developed, the next task is to select a specific mechanical principle.

The Mechanical Principle

What are the mechanics that can make the operational principle actually operational? Although this term refers to the placement, holding, protection, and removal of mail, we will use only placement here as an example. This requires certain mechanisms to provide access: an opening somewhere within the unit. Quite a number of such mechanical principles can be developed: for instance, a front, side, or top opening using a sliding, pivoting, or bending mechanism. Having developed a number of mechanical principles, the task is now to select some of these and develop their technical principles—i.e., to realize them in specific materials and manufacturing processes.

The Technical Principle

The technical principle translates a mechanical principle into an actual piece of hardware by determining the kinds of materials and manufacturing processes that can be used to realize a product. The next task is to relate the obtained solutions to external aspects.

The Relational Principle

Each of the developed operational, mechanical, and technical principles essentially remains tentative until it has been appropriately related to various aspects of the environment in which the unit will be placed and to the persons who will come in contact with it.

Design Problems

The designer is often referred to as a "problem solver," and a design procedure begins with the statement of a problem. Such a statement does not in itself initiate the problem. Although a problem may have existed for some time, it is a problem only if someone becomes aware of it and considers it as such. The threshold of awareness of an existing problem varies considerably from person to person. An individual may accept a certain situation (for instance, an actual malfunction) and simply—perhaps unwittingly—adjust to it, while others may consider adjustment unnecessary and do something about the problem. This is the genesis of a problem and generally occurs in an incidental and spontaneous manner. The question arises: Is it possible to develop some means of drawing attention to and anticipating problems that will probably sooner or later come to the surface?

Referring specifically to the transformer-type object, the kind of problem that the designer has to solve is what we may call an existent environmental "inadequacy"—a malfunction within the environment; specifically, the manner of obtaining certain desired results. As stated before, the transformer-type object is used as part of an operational system to obtain certain desired changes.

The term "inadequate" in reference to objects or environmental aspects means that the object or aspect is considered to be inadequate in relation to the means potentially available at a given time and place. An object can be considered adequate in one place but inadequate in another, or at one time but not at another. Availability of the appropriate means at a given time and place is crucial. There are four areas of possible inadequacy in an operational system. If the designer is aware of these areas, he or she can investigate what inadequacies already exist or can be anticipated in the future. A given object (referred to here as a unit) may have one or a combination of the following inadequacies:

1. Built-in inadequacy due to failure in designing the unit

a. Operational aspects
   - Use of the unit results in unsatisfactory results
   - Too many operational steps are required to obtain the desired result
   - The negative by-results are not acceptable, for instance, fumes or noise

b. Outer constants aspects
   - The unit does not properly relate to or fit the person who uses or otherwise comes in contact with the unit
   - The unit does not properly fit other objects and environmental aspects

c. Temporal aspects
   - The unit as a whole or some of its parts, wears out prematurely

2. Inadequacies, due to changes in the environment in which the unit exists, that have occurred since its introduction

a. Physical and/or visual changes in the environment, directly related to the operation of the unit

b. Physical and/or visual changes in the environment, indirectly related to the unit

c. Changes that have occurred in the prevailing conditions of the environment in which the unit exists

3. Inadequacies due to new technical potentials that have become available

a. New materials

b. New manufacturing processes

c. Increase in the number of units being produced

d. New engineered components

4. Inadequacies due to changes within the person or persons using or otherwise associated with the unit

a. Changes in the physical stature of the user

b. Acquisition of greater skill by the user

c. Acquisition of greater knowledge by the user

d. Changes in attitudes, customs, or behaviour, of individuals, groups, and/or society in general

6

The Ideation Process

In recent years, there has been a rising interest in various aspects of the creative process and in the possibility of consciously influencing the process of ideation. New approaches seeking to facilitate the development of ideas have been introduced. This increase in interest seems to be part of a general trend towards greater awareness regarding the various aspects of the design process.

The following discussion of some of these approaches does not represent a complete survey of what has been done in recent years. It merely serves to indicate the increased interest in this area and perhaps to point to further development. The various approaches are discussed in the approximate order of their historical appearance.

This division of the approach to ideation into stages is the conception of the author and is based on impressions of various techniques in operation. When not explicitly stated by the proponents of a given approach, the basic assumptions have been deduced from the actual teaching and/or designing procedures. (Thus a technique which aims at overcoming "mental blocks" in a student must be based on the assumption that such impediments to creativity exist.) Viewing the development in terms of a series of stages may seem to indicate that a new approach replaced an older one. But, in fact, each new approach is a supplement to all the previous ones. It should therefore be emphasized that many concepts and techniques were, and still are, held concurrently. As in every evolutionary process, there are countless instances here of overlapping, retention, and coalescence. It also should be stated that the historical sequence does not constitute a hierarchy of values. The last development should not be considered as the only valid approach but as an additional method of dealing with the problem of ideation.

My purpose is to demonstrate the status of the major approaches and my view of the current general trend. To achieve this goal, it will not be necessary to trace the history of each approach in detail; only the most characteristic aspects will be discussed.

Let us begin our survey of the attempts to facilitate the creative process with the early man, speculating on the manner in which he first became conscious of ideas existing within him.

In early eras of human development, "ideas" must have been produced with little, if any, conscious effort, as if they had appeared out of thin air. They were probably a spontaneous reaction to problems posed by man's environment. Derived largely from unconscious material, they might well have been considered as coming from outside the individual. This may have been the origin of the notion of "invoking the muse" and other forms of passively waiting for inspiration from external sources. We still cling to the vestiges of this concept when we say: "I got an idea," as if it had been hidden and then discovered.

Only gradually did thinking man become conscious of some sort of internal process involving the production of ideas. As an outgrowth of this awareness, he probably developed the notion that it might be possible to take an active role in producing ideas, thus gaining some measure of control over that phenomenon. This new concept brought forth a more or less formal approach toward the problem of ideation.

The Laissez-Faire Approach

An early attempt to facilitate the development of ideas might be called the "laissez-faire approach." It assumes that the gifted person is born, not made. Talent is regarded as an inherent faculty, which is either present or not present. It is the "divine spark" accorded a chosen few who are predestined to be gifted. In order to foster the spontaneous unfolding of such talent, education should assume the more or less passive role of providing a favorable and protective environment in which the creative faculty will be free to develop according to its own pattern, with a minimum of outside interference.

No provision should be made for persons of "little" or dubious talent, regardless of the effort they might exert to prove themselves worthy of consideration. (An absence of talent is as obvious as its presence.) On the other hand, the truly gifted person may display certain negative personality traits, such as apparent laziness, lack of attention, arrogance, etc. These should be tolerated and treated with the understanding required of an artistic temperament.

Since a totally laissez-faire attitude toward education is self-contradictory and would result in various degrees of bedlam, the main problem is to decide just how much control or guidance is required to avoid "hindering" development, and what form it should assume. The result of these requirements will be a close rapport with the student coupled with a kind of personalized non-directive educational method. Education largely concerns the selection of the "gifted" and the unfolding of the pupil's latent ability to create.

It must have been apparent to the advocates of this approach that even the most "gifted" person is productive only at certain times. Occasionally the individual is aware of the external stimulus responsible for a particular creative impulse, but more often than not an unknown stimulus is involved in the process. This awareness that external stimulation is involved in creativity led to the following:

The Stimulative Approach

This approach also assumes that the talented person is inherently creative. But a more active technique is adopted than the non-interference of the laissez-faire approach. Except in the case of prolific and spontaneous creative individuals (who seem to thrive under any circumstances), a dormant or more or less quiescent talent is thought to exist. The function of this method is to single out the "talented" person and bring latent creativity to the surface through application of, or exposure to, a particular stimulus or set of stimuli.

The forms of stimuli are as many and varied as are the requirements of different individuals. The unique psycho-physiological make-up of each person predisposes him or her towards certain kinds of stimuli and renders others ineffective. This student-stimuli relationship calls for continual adjustment and flexibility on the part of the teacher. Thus the study of the humanities, the physical sciences, and the social sciences, may be capable of providing indirect stimulative experiences. Other forms of stimulation, to name only a few which are used in the classroom, are the effect made by a teacher with a vital presentation, discussion, approval, opposition, criticism, team-work, and competition. In short, it can be said that stimulation occurs where YOU find it.

In essence, the two preceding methods are of great antiquity and have probably existed in one form or another since the beginning of recorded history. In the last few decades we find a renewed interest in the subject, the German Bauhaus being one of the main focal points. It was here that the following approach first appeared.

The Liberation Approach

This approach is similar to the preceding ones in that it assumes that creativity is an inherent faculty. But it adds another dimension—a more tolerant attitude toward innate design ability. Maholy-Nagy summed this approach up in the following statement: "The basic idea of this education is that everybody is talented—that once the elementary course has brought all his power into activity, every student will be able to do creative work."[35] However, the proponents of this approach recognize that an individual's ability to create is often inhibited or blocked by internal and/or external forces, which partially or completely impede its realization. Therefore a process of "unlearning" or generally "loosening up," was considered essential to further development. The elementary or preliminary courses, variously called the Basic or Foundation Workshop in the Bauhaus literature (or Vorkurs 2 at the Bauhaus) served this function (see my critical review of the Bauhaus course, in Chapter 1). Here the students were encouraged to manipulate a great variety of materials and tools in both conventional and unconventional ways. They developed and "experimented" with new forms, structures, textures, relationships, etc. Through these exercises in spontaneous manipulation where "the sky is the limit" coupled with individual and group discussions, it was hoped that the students would overcome their design inhibitions and be able to develop freely without the hampering effect of previous attitudes.

It was assumed that the freedom of approach acquired in the preliminary workshop would be carried over into the area of practical application and result in the production of new products. The function of education was therefore "to help [the student] overcome self-conscious fear and thereby learn to employ his native ingenuity without restraint..."[36]

The Counteractive Approach

This approach also seems to have originated in the Bauhaus movement. There are striking similarities in the basic assumption of both—that mental blockage can impede creativity. In the liberating approach of the Bauhaus, more emphasis is placed on internal (psychological) inhibitory factors than on the primarily external conditioning forces with which the counteractive approach is concerned. While taking into consideration problems of personal disposition, this approach also concerns itself with the conditioning process resulting from continual exposure to the customs and conventions of our culture. In the routine of daily existence, our thought patterns tend to become canalized along familiar routes, shackled to conventional ways of thinking and solving problems. From a creative point of view, a mind so conditioned can utilize no more than a fraction of its potential and, in attempting to develop new ideas, will find itself thwarted by an inability to break away from previous conditioning patterns.

Based on the premise that a familiar stimulus tends to evoke a familiar response, this approach has been developed to counteract such a tendency. To this end, assignments are set up which force the student to abandon his usual everyday thought pattern. For instance, the student is asked to design implements for a race of people having no opposing thumbs; the problem is stripped of most preconceptions and the student is forced to exercise his brain in a different way. Similarly, in an assignment to design a shelter for a race of quadrupedal individuals, there is little chance for conventional handling (introduced by Albert Szabo, at the Institute of Design, Chicago, 1950). Such a setting up of artificial situations was carried a step further by the late John Arnold, who invented a whole new "world" in order to stimulate original and creative thought patterns in his students. Thus his Arcturus IV and Ceres projects present the possibility of designing objects in a new dimension, for a race of people who differ radically from our own.[37]

The Brainstorm Approach

This approach was developed by men in industry faced with the task of solving problems in their field.[38] It was subsequently applied to design problems. The basic premise of this approach seems to be that, in life situations, fear of criticism often hinders the production of new ideas and results in a tendency to "play safe." Therefore, if criticism could be withheld and attention given to ideas as such, new solutions would come to light. Thus the approach requires the creation of a psychological atmosphere in which a free exchange of ideas can take place. This occurs in arranged meetings or sessions of small groups interested in solving a particular problem. The individual participants do not have to be experts in the field or otherwise familiar with the problem at hand. The problem is stated very specifically and each participant offers his or her solution without restraint. Since the primary goal is maximum spontaneity of expression, there are only a few general rules: a closed session never extended to become tiresome; a limited number of participants; and the most basic rule of all, no passing of judgment in any form on suggested solutions.

Each participant is encouraged to express ideas freely without fear of encountering criticism, ridicule, or any disapprobation from other participants. Ideally, this setup produces a psychological atmosphere which facilitates reciprocal group stimulation and a "chain reaction" form of ideation. In a relatively short time, a great variety of ideas are brought to light for later analysis and evaluation.

The Checklist Approach

While the use of the checklist does not actually constitute an "approach" to design, it is included with the above-mentioned facilitative approaches because I believe that in the future it, too, will become an important component of the conscious approach to ideation.

The increased use of the checklist is indicative of the designer's awareness of the increasing complexity of the designing process. Since it is no longer possible to keep mental tabs on all the factors involved in a design procedure, the listing of all pertinent data which must be considered in the designing of an object has become an indispensable tool in the solving of design problems.

Today's checklist, which includes the number of steps in a procedure, their order, and the relative importance of each step, is essentially the result of personal preference and experience and is commonly developed for personal use.

Experience, combined with the new requirements, will modify, refine, and improve the existing checklist, until a stage is reached where it will be possible to develop it on a logical, non-personal basis. The present writing constitutes a step in that direction.

The above survey indicates a development from spontaneous creative behavior towards the use of a more or less specific approach; from a fear of tampering with the "natural" creative process to a conscious attempt to develop a means of facilitating creativity. We are no longer merely designing; we are actually designing the designing process.

Exercises in Ideation

Exercise A: The "Space Modulator" Exercise

This exercise was originally developed at the German Bauhaus by Moholy-Nagy during my second semester under him, and it since has been adopted in various schools. The reason for referring to it and other assignments here is to draw attention to the fact that certain factors of which the designer may by unaware sometimes influence the design process.

The construction of a so-called space modulator was assigned in a course designed to further the development of the students' creative ability. It is perhaps better referred to as a free-hanging sculpture. In a class which had a rather large number of students, three groups were formed. Each had approximately thirty students. The three groups were scheduled to work at different times but in the same room. The problem of storage of the students' work was solved by hanging it on wires strung from wall to wall below the ceiling. Toward the end of the assignment the ceiling was festooned with a complex maze of space modulators. Viewing the mass of work, one could observe certain general characteristics, which revealed interesting facts about specific configurations.

The usual material used in this exercise was wire interspersed with plastic, cardboard, and other materials. The wire was supplied in the commonly sold coiled form. The student cut off as much wire as required and proceeded to make a space modulator. This program went on for a number of semesters. Once, when new material had to be ordered, the instructor requested the same type of wire in straight lengths of approximately ten to twelve feet instead of the usual coil.

When the next semester began, the students were given the same assignment as in the preceding semester. The work in progress again was hung below the ceiling; and by the end of the assignment, the ceiling once more was a maze of space modulators. But this time, the general picture was quite different. Whereas the general appearance of the work made by the students of the preceding semester, using coiled wire, was almost exclusively curvilinear; the work resulting from the use of straight wire was predominantly straight and rectangular. There were still some curves—just as, in the preceding groups, there were some straight lines and angles—but the dominant form for each group was clearly different: curves dominated in the group using the coiled wire, while straight lines dominated the group using straight wire.

Exercise B: The Paperfold Assignment

This assignment was intended to further the awareness of factors which may influence creative activity.

The students were given a sheet of paper measuring approximately 8 ½" x 11" and were asked simply to make a fold in any form or manner. The kinds of folds that would be produced were to some extent predictable. All would be straight, not only because the students would assume that folding paper will result in a straight fold that bends the paper back on itself, but also because a straight fold is easily made, in contrast to a curved fold. Since the various folds produced by the students were in accordance with the commonly accepted notion of straight folds, they generally had certain characteristics in common. Most were related in some way to the original rectangular shape of the paper—an aspect that had no particular relevance to the assignment, since no reference was made to this. Actually, any completely arbitrary fold would have fulfilled the requirements of this assignment. However, students seem to have an aversion to such "arbitrary" action. They tend to justify their folds by relating them, in one way or another, to the original shape of the paper. This may be the reason why some folds were precisely made in the center, requiring a simple decision since there are only two such possibilities. Because there are quite a number of other possible relationships—i.e., one corner related diagonally to the opposite corner, a short side to a long side, etc.—one can easily "justify" the position of the folds by referring to these specific relationships. The folds made by the students were seldom "arbitrary;" that is, without some apparent relationship to the form of the sheet. One could expect similar results if one were provided with a triangular piece of paper, a free-form, or even a torn sheet. The important point is that most of the results of the paper-folding exercises were directly influenced by the original form of the paper, as with the wire in the space-modulator assignment.

Exercise C: The Dot Assignment

In another assignment intended to increase the students' awareness of their creative potential, a sheet of paper was furnished which had four, barely visible dots. No reference was made to these dots; they seemed to be merely incidental. The assignment was simple: "Make four lines with a pencil in any shape or manner." The results were then collected.

In the next assignment the student was again given a sheet of paper. This time there were two very faint horizontal lines on it. The assignment was to make five lines. Again, no reference was made to the marked lines. After these sheets had been collected, the student was given a third sheet of paper with a few freely made, but only faintly visible, wavy lines. Now the students had to draw six lines in any way they pleased. All the results were then displayed and discussed. Basically, these assignments were the same except for the number of lines. The intention was to draw attention away from the fact that all the assignments were fundamentally the same. So far as the assignments were concerned the students could have made the same number of lines anywhere on a sheet. They could have made a square, a pentagon, a hexagon, or any other line formation. Nevertheless, we found that most of the students' lines were in one way or another related to the faint markings on the paper. In the first two assignments, the lines not only showed such a relationship but were almost exclusively straight. The third assignment produced quite different results. The lines were rather long, wavy, and freely made. There was no restraint, as in the preceding assignments. The goal of these exercises was to demonstrate clearly the influences to which a creative person is subject.

As part of this assignment, and in order to draw further attention to the influences on the creative process, the following hypothetical case was described: someone is asked to make a drawing which includes a number of long parallel, but slightly curved, lines. A difficulty

fig. 25 (opp. page, left)
Dovetail Joint 1
fig. 26 (opp. page, right)
Dovetail Joint 2

108

emerges since the making of a curved line requires different tools than those used in making a straight line. Although the difficulty may be minor, it may still evoke the question whether the lines had to be so slightly curved. The thought arises, could the lines just as well be straight? The readily available straightedge could then be used. There is nothing wrong with such critical thinking. However, the important point is that the question generally arises only because of the difficulty in making the curved lines. In contrast, a straight line might not be questioned, because it is so simple to make and is so readily taken for granted.

In the stated examples, the influences are easily recognized as soon as they are pointed out. However, one must assume the presence of other more subtle influences which are not so obvious. These influences are manifold and may be due to the tools, the materials, the media being used, or the skills, tendencies, and inclinations one possesses or lacks. The important point is that one should be aware of such influences in order to make a deliberate choice—to choose to use or not to use whatever comes to mind.

Exercise D: The John Arnold Demonstration

The late John Arnold used to talk about the following demonstration: Approximately in the center of a stage setting, there is a not particularly attractive 30-inch long sewer pipe embedded in concrete in a vertical position. At the left of the stage is a workbench with a three-foot long stick and a common blacksmith's tongue. At the right of the stage, there is a rather fancy mahogany table and a silver pitcher with water on an elegant piece of lace. In this setting, the instructor drops a ping pong ball in the sewer pipe. The problem is to recover the ball. The pipe cannot be turned over, nor can one reach to the bottom of the pipe. The thought may occur that the stick might be useful, but this method would be difficult. The blacksmith's tongue could hold the ball, but the narrowness of the pipe would prevent the tongue from being opened far enough to grab the ball. Of course, there is the pitcher full of water, and pouring the water into the pipe would bring the ball to the top quite effortlessly.

The point of the experiment is that the general tendency is to try first the stick or the tongue, because they are more readily associated with a sewer pipe, particularly if the pipe is not

very clean. Only after the stick and the tongue prove to be of no use will the silver pitcher, which is not readily associated with a sewer pipe, be resorted to. The intention of the experiment is to make the observer aware of his or her own inclinations and tendencies.

Exercise E: The Dovetail Extension Joint

There is something baffling about the illustrated dovetail joint (fig. 25). The question arises, how is it put together? It would appear to be an impossibility. (An individual with little structural understanding or curiosity would not be too concerned.) Compare this joint with another (fig. 26). The second one is more comprehensible. What is the difference between the two? There actually is none—internally they are identical. However, there is a difference in their outer form. The second joint could easily be made to look precisely like the first one. All that is needed is to cut the corners at a 45-degree angle. The first joint was actually made in this manner. The baffling effect is caused by our conclusion that the inner structure follows that of the outer form; that is, it must be parallel to the sides of the outer form of the unit. Indeed, we frequently have to deduce the inner structure of an object from the appearance of its outer form.

Various aspects of a design procedure tend to influence or lead the designer's creative efforts. These influences are referred to here as the "lead" phenomena. Only an awareness of the potential influences will give one the choice whether to follow or not to follow such a lead.

The Classification Approach to the Development of Ideas

The classification approach is based on the fact that, in classifying some aspects of an existing sketch or product—i.e., in referring to one specific characteristic—attention is drawn to other characteristics not present. In other words, reference to a specific characteristic places the object in a particular class with other objects bearing the same characteristic. Such a class exists only to the exclusion of other classes. This implies that "if I know what I have, I also know what I do not have."

The following assignment—to develop a circumscribed area graphically—will serve to illustrate this. Let us assume that one begins to make a sketch of an elongated parallelogram. By stating its obvious characteristic, one places it in a group of parallelograms. To begin with, it should be pointed out that this is not the only characteristic one could refer to. Even such a simple shape can be classified in a number of ways (more will be said about this later). An awareness of this fact suggests not only that there are other parallelograms, but also that there are shapes that are not parallelograms. Thus, a new class of non-parallelogram shapes is developed. Further development can now move in three directions. First, each of the two established groups can be developed further. Generally, however, it will be more profitable to think of a completely new characteristic, other than presented by the two groups. For instance, both groups have regular shapes. This suggests the development of irregular shapes. Now we have three groups. And again, in establishing the common characteristics of these three groups, other new shapes are suggested. For instance, all are made with straight lines. The results can again be dealt with in the same manner, producing new shapes. The development can always proceed in two directions: (1) extending already established groups further; and (2) classifying all the ideas which have arisen to establish new groups. What do ALL of the forms produced so far have in common? How far should this procedure go is a matter of available time and the ability to classify further?

Two arise: What kind of classification is involved? And, since the above illustration is restricted to graphics, Does it apply equally to the development of products? Apropos of the question of classification, the potential number of such simple graphic shapes seems to be rather limited. It is nevertheless greater than initially appears to be the case. We have already dealt with various kinds of classification in a previous chapter. These can now also be applied to the development of ideas. For instance, the three basic attributes of entities (structure, form, and position) can be used to create new, essentially different, groups. If we notice that certain shapes which have been developed take into account only the attribute form and ignore structure and position, this reminds us that those shapes can assume various positions and that their structure can be varied as well. Another direction might involve the realization that specific shapes were created by using a line-producing medium. This happens to be a convenient and readily available medium. Thus a surface-producing medium might be suggested.

What has been said about graphic assignments applies also to products. For instance, in reference to the operational, mechanical, technical, and relational aspects of a product, a sub-grouping of the operational aspect could list the order and number of steps required to obtain various results.

It will not, as I have already said, be possible to deal more extensively here with the different aspects of this approach.

Personal Aspects Affecting the Creative Process

In general, man displays considerable awe and respect for the creative process, often to the extent of not even attempting any creative activities because such activities are assumed to be reserved for special people. Although everyone has to cope with problems, these are generally of a personal nature. Designers, however, have to deal with the problems of society; and society will sooner or later demand that they find a solution. The individual designer merely gets a chance, an opportunity; and if he or she fails to find an appropriate solution or ignores the problem altogether, someone else will take over the task. In solving a problem, it is not simply a matter of obtaining a solution but of obtaining an optimum solution (the optimum possible in a given space-time continuum). Although the problems the designer has to cope with are those of society, they have to be dealt with by someone whose personal makeup will influence the design. The designer's personal inclinations and tendencies, particular interests, and likes and dislikes will play a role in the creative process, even though the designer may be unaware of, or have an "I don't care" attitude regarding that particular makeup. If the designer seeks to go beyond mere personal satisfaction, it will be necessary for him or her to be aware of these personal tendencies.

The following listing may help to draw attention to such traits.

• Skepticism regarding preconceived solutions. Most likely they represent only one of many possible solutions.

• Recognition that an idea in itself (particularly if it seems to be a good one) can "blind" or prevent us from searching further. There may be a tendency, in our eagerness to find an appropriate solution, to latch on to an idea that does not represent the optimum possible at the time. The evaluation of a particular solution requires others for comparison.

• Giving up on an idea too quickly, just because it presents a difficulty in the solution of a particular aspect. One may have a tendency to merely touch upon an idea and then discard it immediately, rationalizing that "it is too complicated," or "it won't work anyhow," or "it will be too expensive." This may simply be a cover up of a personal aversion to sticking to an idea and discovering its actual potential. (It should be remembered that the first aluminum product was prohibitively expensive, and that it was once thought the automobile would never succeed because it would be too expensive to run.) It is almost a general rule that the "new" requires special mental and physical effort.

• Retaining a sense of proportion toward the whole problem. Excessive concentration on particular details puts the designer at a disadvantage.

• "Presence of mind." Imagine yourself in the situation in which the problem exists.

• Changing the "frame of reference." Anticipating how different people will act in such a situation.

• Oscillation. Being intimately involved yet infinitely removed (W. Gorden).

• Maintaining a critical attitude about all aspects of our environment, without being constantly in a frustrated state of mind. This means that one should oscillate between being critical and not critical; i.e., between accepting and not accepting aspects of the environment.

• Knowing your own personal tendencies and inclinations. Awareness of the extent of those tendencies; for instance, being too frugal or excessive in your way of living, which in turn may affect your solution.

• Ignoring the notion of self-expression. Expressing oneself in a merely incidental manner, in the way one goes about designing a product.

Developing the mind

When we refer to the results of a designing or inventing process, we often talk of having "discovered" something. This seems to imply that something was hidden from us. We tend to refer to a discovery as something outside (not within) us; since it was difficult to "find" it must have been hidden. Actually, the fact is that nothing is hidden and therefore nothing can be "discovered." The trouble or difficulty encountered in designing or inventing something comes from within. The matter and entities from which all man-made inventions derive have existed since the beginning of time. What was lacking was the capacity to understand and synthesize existing matter or entities in order to produce a design or invention.

In the process of designing and inventing, we are actually training and developing our inner selves—our minds. The matter and entities we are physically and intellectually manipulating in this process are the means by which we are being transformed. The knowledge we accumulate and the theories we develop are not only the outward sign of an evolving mind but the very means for bringing about that evolution. Just as one tool can be used to make another more advanced tool, so a searching mind can become the catalyst for further advances. All advances in the development of our man-made environment depend upon the development of the mind, both on an individual and on a collective basis.

7

Education of the Designer

The general educational program for the industrial designer includes a number of academic courses. However important they may be for the student in general education, they will not be dealt with here, since our focus is on design courses. In most schools, these are essentially laboratory courses, in which the student spends a good deal of time solving a variety of design problems. The approach is somewhat similar to what we find in general practice. It is essentially an apprentice-type approach: learning by doing, training rather than educating.

The task of design education is of course to prepare the student realistically for the profession with all its complexity. But while education is needed to practice design, a facsimile of design practice is not necessarily an adequate educational approach. The difference is that while in practice the prime concern is for the end-result—i.e., obtaining a viable product—in education it is for the process that results in a product. It is important for the student to become aware of the various aspects and phases of a design process that can be applied later in the designing of various kinds of products. Thus, the educational program presented here emphasizes knowledge conveyed through a series of lectures.

The Two Design Course Series

The two design course series which I introduced—the "Situation Course Series" and the "Application Course Series"—were developed in order to deal appropriately with the two ways in which people create products. Each of these series followed a somewhat different approach. Since the approaches have been dealt with at length in Chapter 6, they will only be recalled here briefly. The Situation Course Series starts with the statement and definition of a given situation and proceeds with the development of the means applicable in that situation. The Application Course Series starts with a given material, entity, and/or manufacturing process and seeks to apply it in an appropriate situation. In both course series, the student solves a variety of design problems.

The assignments given to the student in these courses serve a somewhat different function than is customary. In the course series, each assignment is accompanied by a lecture covering specific aspects of the man-made object. Thus the assignments serve essentially to give a vivid understanding of the practical application of the knowledge dispensed. In the Application Course Series each assignment also serves to familiarize the student with major materials and manufacturing processes and, importantly, with their appropriate applications. Beyond this, each assignment also serves to develop the student's skill and technical training, as is illustrated further in the following discussion of specific assignments

About Assignments

An important part of design education is the assignment of problems. The kinds of problems vary widely in different schools; what may be considered important in one may be ignored in others. How are these problems selected? We have no general survey covering major schools of design, so we have to rely on publications, exhibitions, and occasionally visits to schools for an answer.

Based on available information, a consistent assignment program seems to be lacking. The selection of design problems is apparently dependent on the personal interest and inclination of the instructor; and whatever kind of assignment program may exist; it is often interrupted by various awards programs staged by industry and institutions. Basically, there is nothing wrong with a school participating in such a program, provided that it accords with any assignment program the school may have. The following outline is an initial attempt—a possible basis for further discussion—to establish a consistent assignment program that would adequately cover all major aspects of the process of product design.

The Situation Course Series: Outline for an Assignment Program

Based on a general study of the man-made object and the process by which it is produced, an assignment program for the Situation Design Course Series would involve the following problems, assigned to students approximately in the order in which they are listed below. In all of these assignments the emphasis is on the process, on awareness of the different steps involved in applying the knowledge gained from the accompanying lectures, and on solving a variety of design problems.

The Fifteen Problems addressed in the
Situation Design Course Series are as follows:

Problem 1:   The ideation process:
             an introduction to the development of ideas

Problem 2:   Form as a relational quality

Problem 3:   The inner build-up of a form

Problem 4:   The surfaces of forms

Problem 5:   The visual relationship of a form

Problem 6:   The means employed to produce form,
             and their influence on form

Problem 7:   The graphics in a design procedure

Problem 8:   The basic principles of an object

Problem 9:   Designing structure:
             the inner build-up of objects

Problem 10: Designing special structures:
             changeable structures and access

Problem 11: Anthropometry:
             the relationship of objects to users

Problem 12: The analytic approach to design

Problem 13: Designing products with engineered components

Problem 14: The object as a detail of the environment:
             ecological aspects of objects

Problem 15: The development of objects over a period of time:
             the temporal aspects of objects

fig. 27 (opp. page)
Development of Initial Emblem
in Order to Demonstrate
the Process of Ideation

Problem 1: The Logo Assignment

Interviews with students planning to enter an industrial design program reveal that many have doubts about their ability to produce ideas. They seem to be concerned whether they will be able to develop the quality and consistency demanded by industry. Such fears can probably be alleviated by providing students with knowledge and an understanding of the specific aspects with which they will be confronted.

To familiarize students with an approach to the generation of ideas, they are assigned the development of a logo using their own initials. For simplification, only two initials are used. It should be emphasized, however, that something other than initials—e.g., an object—could be used just as well. The important point is that the subject of the assignment should be simple in order to keep the emphasis on the process rather than the end-product. Using graphics as an introduction to the development of ideas has the advantage of avoiding all technical considerations at this point. Moreover, while the assignment is basically the same for all students, it varies from one to the other. The assignment seems to have an added interest for students because many of them have already taken some interest in their own initials. Although the emphasis is on the process, the quality of the final result is not ignored.

The aim of the assignment is not to develop one "good" logo but rather to generate a great variety of different logos, and to do so in a deliberate manner. This helps the student to overcome the notions of "my" design, "self-expression," or "falling in love" with a single idea, all of which hinder further development.

The procedure followed in this assignment is the "classification" or "grouping" approach. Lectures and demonstrations introduce the student to this approach. Although it has been discussed at length in an earlier chapter, it will be briefly repeated here:

The student begins by sketching any initial that comes to mind, disregarding its "good" or "mediocre" quality as long as it is relevant to the assignment. Under no circumstances does the student wait for the "brilliant" idea. At this stage, sketches do not constitute a commitment or give any indication of ability. Through analysis—i.e., characterization and classification—

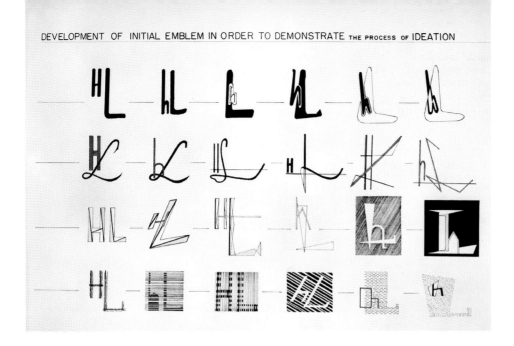

DEVELOPMENT OF INITIAL EMBLEM IN ORDER TO DEMONSTRATE THE PROCESS OF IDEATION

these first sketches serve to develop new ideas. Characterizing a given form means classifying it—i.e., putting it in a class of objects having the same characteristics. The implication of such a classification is that there are other classes which do not have these characteristics. These can now be developed. The results—either of a single logo design or of a group of designs— are again and again subjected to this treatment. It seems to be that only the limitation of time determines how long the exercise should be continued. The final presentation does not merely show one preferred design, but illustrates a number of widely different groupings that have been obtained (fig. 27).

Problem 2: The Physical Contact Relationship Assignment

This assignment considers the relational quality of an object's form. It is a further development of the so-called hand-sculpture assignment I introduced at the New Bauhaus in Chicago in 1937. My intention in developing the hand-sculpture assignment was to draw attention not only to the visual but also to the physical relationship of man to object as an important factor determining form. In the years that followed, it became apparent that the making of a hand sculpture (like the creation of a tactile chart) merely drew attention to an important aspect of an object without providing any further information. It did not show the student how to apply the experience gained to the designing of a product. For this reason, the assignment was changed; rather than making an isolated hand sculpture, the student had to determine how and to what extent an object's form is shaped by man's physical relationship to various aspects of the environment. Reference was no longer to single objects but to man's environment in general. Although the assignment merely calls for the development of a simple object, it is meant to underscore and to make vivid man's physical relationship to environmental aspects and their influence upon the forms of our objects. These relationships are extensively dealt with in an accompanying lecture.

120

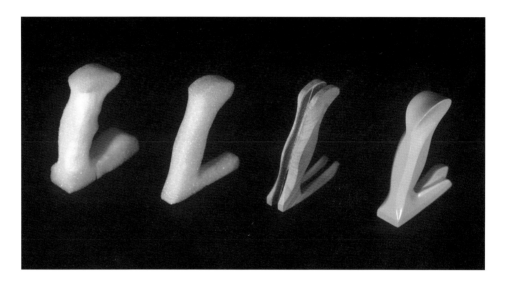

A specific assignment might call, for instance, for the development of a handgrip for a tool or utensil. In developing such a handgrip, the initial emphasis is on the physical relationship, followed by a concern for the visual relationship. A handle that fits a particular person to perfection could easily be made by casting a form directly from the hand of that person, thus obtaining what we call a form-fit handle. This could then be modified to fit a range of persons, changing the form-fit to a variable-fit (fig. 28). The resulting form, however, may not be acceptable from a visual point of view. Since an object is experienced not only when it is used but also when it is not in use—i.e., essentially visually—this aspect of a form has to be taken into consideration. The approach to the visual aspect in this assignment can be illustrated through three types of handles designed by students, each representing different characteristics from a visual point of view. Type A (fig. 29) includes handles which fit the hand. The forms are all smooth and well rounded but do not appeal visually. They might be called "formless" (if such a term makes sense). Our eyes seem to slide over the form without being able to get hold of it. Essentially, we are seeing only the silhouette of the form and are missing what is within the borderlines. There is a sense of uncertainty. The Type B handles (fig. 30), which not only fit

the hand but also have visual appeal, are very different. The form is defined, and one can experience the whole form, not only the silhouette. It appears controlled and thus represents a finished product. The Type C handle (fig. 31) represents an attempt to overcome the "formlessness" of the first type, but the designer has not succeeded in obtaining the finished form represented by Type B.

Problem 3: The Inner Build-Up of a Form Assignment

The previous assignment dealt with an object's form but ignored its inner build-up. In the process of developing the first handle, the student's thinking, as well as the actual making of the object, was based on a method of subtraction. In shaping a piece of wood, the inner build-up is merely incidental; it is design by default. To draw attention to this fact, the next assignment requires the student to make another handle (precisely the same type made before) based on the inner build-up of the form. This is a process of adding instead of subtracting, of more precisely placing the material where it is actually required (fig. 32).

As stated before, this and the preceding assignment actually deal with the general subject of man's physical relationship to aspects of the environment even though it merely calls for the development of a handle for a tool or utensil. When we look at our man-made environment,

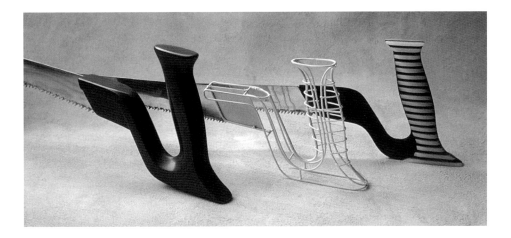

fig. 33 (opp. page)
Tactile Chart Assignment

we see clearly that the rectangular form is dominant. This form is not necessarily derived through a direct relationship to a person or persons. Other factors account for much of the existing squareness in our surroundings—for example, such productive and economic factors as the need to obtain maximum closeness in storing, shipping, and utilization and to save space and prevent movement. There are environmental situations, however, in which square shapes are not justified and where it would be better to relate the form to the persons using the unit. To illustrate such a situation, let us picture a building on a street corner with a lawn between the sidewalk and the building. It was the designer's intention that a pedestrian would make a sharp right-angle turn at the corner, reminiscent of a marching army. However, some pedestrians would find this absurd and would instead walk across the lawn. The questions arise: Is the pedestrian violating the intention of the designer? Or did the designer fail to facilitate the behavior of the pedestrian? Numerous cases can be cited in which the designer fails, at least in some respects, to facilitate the requirements of the user. In the case mentioned above, a user can, if so inclined, simply ignore the designer's intention by walking across the lawn, thus "correcting" the design to some extent. There are other situations (or products) in which the user has no choice but to accept the situation as it is and is forced to make a personal adjustment. Some users may actually be unaware of doing this; others may feel discomfort and aggravation.

Although everyone has to make some adjustments when using a product or an environmental aspect (this was discussed in Chapter 5), these adjustments should be no more than currently available means demand. For it is the task of the designer to facilitate the demands of the user by exploiting the available means.

Problem 4: The Surfaces of Objects Assignment

This assignment is a further development of the so-called tactile chart introduced by Moholy, first at the Bauhaus in Germany, and later in 1937 at the New Bauhaus in Chicago. A tactile chart consists of different textured materials selected and arranged by the student in some fashion. As a student at the Bauhaus in Germany and later as an instructor at the New Bauhaus in Chicago, it became clear to me that there is actually little educational benefit in making a

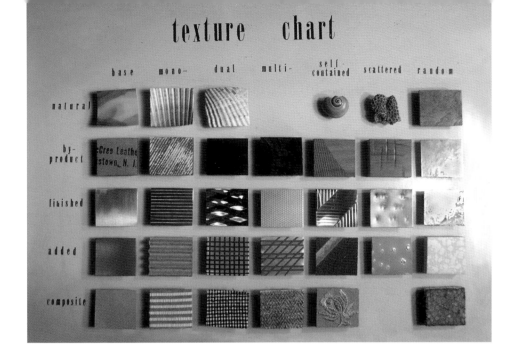

texture chart

	base	mono-	dual	multi-	self-contained	scattered	random

natural

by-product

finished

added

composite

tactile chart. It merely draws attention to the texture of surfaces without providing any information applicable in the designing of products.

Although a surface is a sub-attribute of a form, the preceding assignment did not pay any particular attention to the surfaces of the forms produced. This assignment, instead of calling for the making of a tactile chart, the mere arrangement of different textures according to the student's inclination, introduces a general study of surfaces and surface treatments. It includes the classification of different kinds of textures and texture-patterns, the origin and production of textures, and their application in the designing of products. Students do not repeat the assignments of a preceding group. Collectively, they seek to advance our understanding of surfaces and surface treatments and their role in the designing of products (fig. 33; see also Chapter 4).

Problem 5: Visual Relationships Assignment

In all of the preceding assignments some visual aspect is involved but only as a secondary consideration. In this assignment, the visual aspect is the primary concern. Specifically, the assignment calls for the shaping of a piece of material for evaluation from a visual point of view. Defined in this manner, where "the sky is the limit," most students find the assignment too vague. In order to avoid loss of time while the student searches for a way to approach the problem and a place to begin, he or she is furnished with a piece of wood, approximately 2"x 2"x18". Because of the original shape of the material, there will be a certain similarity in the works of different students. The results will probably all have an elongated shape. Nevertheless, sufficient freedom remains so that the student can deal adequately with the intended purpose. The task is not to arbitrarily shape the piece of wood but to create a recognizable order. Within this framework, the student is free to create any form. Any available tool can be used. To avoid the notion of "self-expression," the resulting work is not referred to as a "sculpture" but simply as a "VR" (visual relationship). As in other assignments, the student is not simply proclaiming "that's the way I like it," but rather to a specific order of the form created. Here, like in other assignments, the student expresses his or her individuality in the manner of approaching the task (fig. 34).

Problem 6: The "Woodcut" Assignment

This assignment was first introduced by me in 1937 at the New Bauhaus in Chicago. Its first aim is to make the student aware of the influence of manufacturing processes on the forms of our products. Secondly, it requires the student to look for potential forming possibilities in a given

process that can be exploited in the designing of a product. Whatever the result of a design procedure, the product has to be produced by a machine. This is not always a routine matter; a possible feedback from a manufacturing process to the design process ought to be considered.

Each manufacturing process leaves a typical imprint on the product. Some imprints are minor; others actually dominate the form of the product (for instance, the potter's wheel). It is up to the designer to accept or reject, in a deliberated manner, whatever these imprints are.

The forms produced by the student in this assignment are not necessarily applicable, as such, in the designing of a product. The intention of this assignment is to draw attention to possible influences of manufacturing processes on the form of a product and hidden potentials of a manufacturing process. This aspect is further dealt with in the Application Design Course Series section.

After the students have been introduced to the known forming possibilities of a given machine, their task is, not merely to repeat the performance, but to go beyond what is known and to experiment freely by producing new forms. The work is to be made completely by machine with no handicraft is involved. The use of wood (hence the term "woodcut" assignment) and a particular machine is merely incidental in this assignment. Any material and/or process that happens to be handy could be used just as well (fig. 35).

Problems 7 and 8: The Manual Presentation of Concepts and the Principles of a Transformer type of Object

This twofold assignment introduces the student to the different kinds of graphics generally used in the designing of objects. These include notation, realization, variation, optimization, and visualization (see Chapter 5). Simultaneously, it familiarizes the student with the different principles of a transformer type of object: operational, mechanical, technical, and relational (see Chapter 2). Since both aspects have been dealt with at length in a preceding chapter, it will not be necessary to do more than mention them here. By now the student should have acquired sufficient skills and knowledge in sketching and various kinds of drawing in previous courses and be able to apply these in an actual design procedure.

The essence of this assignment goes beyond the mere designing of an object. It is more important that students acquire an awareness of the influence of their own make-up and tendencies in designing. The student, for instance, may be prone to making endless notation sketches, either waiting for a "brilliant" idea or finding it too inconvenient to go on to the next phase of the procedure that may demand different graphics and a different mental effort. Or the student may concentrate on only one of the principles of an object, neglecting the others.

The object that the student develops should be simple, so that emphasis can be placed on procedure rather than on the final product. A mailbox, breadbox, birdfeeder, or similarly simple object could be used in this assignment. For further details, see Chapter 6.

Problem 9a: Designing Structure Assignment

The forms developed in the preceding assignments are made without particular reference to structure, the inner build-up of a form. The only exception is the second type of handle in which the inner build-up was considered. This assignment deals specifically with the designing of

structures. A chair would be an appropriate object for demonstrating the difference between designing structure and designing form. In the designing of a chair the form is, of course, of vital importance. It is relatively simple to fit a chair properly to a person, essentially a variable contact relationship. Beyond this, the designing of a chair is primarily a structural and technical problem. The questions are, how and by what means can one support a body in a certain position? In other words, one must develop specific structural members and place them in strategic and/ or formal positions.

The task requires three-dimensional thinking. However, when thinking in terms of spatial structures, the mind seems easily confused because it has a tendency to think in two dimensions, on one plane. To overcome this handicap the student is encouraged to work not only graphically but in three dimensions, using wire, sticks, dowels, etc. This makes it possible to sketch in space, working in scale or with full-size models (fig. 36). The objective is to obtain a structural principle appropriate for a certain type of chair. When a number of suitable structural principles have been developed, one is selected; and its various dimensions and appropriate materials are determined so that a prototype can be made. At this stage, drawings are kept to a minimum, just sufficient to produce a prototype. This helps the student to retain a creative attitude in all phases of the procedure and to be willing to make changes if they are called for. Final drawings and possible renderings are made only after the execution of a prototype.

To encourage students to think beyond the one particular structure they have produced, a parallel assignment calls for the designing of a series of chairs, each having a different structure. The students are presented with a list of basic structures suitable for chairs. They then design (in sketch only) a series of chairs. All must be of the same type—simple, straight chairs—keeping this form essentially constant. The aim is to arrive at a pure representative of each of the different structures, so far as this is feasible (fig. 37).

fig. 38 (opp. page)
Beanbags Illustrating
Granular Structure

Problem 9b: A Grouping of Different Structures Suitable for Chairs Assignment

The following list is an initial attempt to establish a grouping of possible structures suitable for chairs. Since the grouping is tentative, it is presented to the students as a challenge, encouraging them to question the grouping and perhaps suggest further improvements. The underlying aim here, as in other assignments, is to encourage the student to think in broad terms, beyond a single design solution.

1. Compositional structure: the entire inner build-up of the form consists of one material.

2. Skeleton structure: the inner build-up of the form consists of a number of structural members.

a. Skeleton-space structure: structural members are within the space confined by the form.

b. Skeleton-contour structure: structural members follow the contour of the form.

3. Shell structure: the inner build-up of the form consists of sheet material.

a. Sheet-space structure: the sheets are within the space confined by the form.

b. Sheet-contour structure: the sheets follow the contour of the form.

4. Pneumatic structure: the form is maintained by inner air pressure.

5. Hydraulic structure: the form is maintained by inner water pressure.

6. Granular structure: the form is maintained by inner granular material (fig. 38).

Problem 10a: Designing Changeable Structures Assignment

Whereas the preceding assignments call for the designing of rigid structures, this assignment considers structures that can be changed in order to adapt their forms to different situations or persons. There are a number of reasons for changing an object's form. A form may have a changeable structure so that it can be adapted to different:

1. persons: changing in accordance with different statures and weights,

2. situations: changing from a use-position to a dormant-position, or from a shipping-position to a storage-position and use-position, and

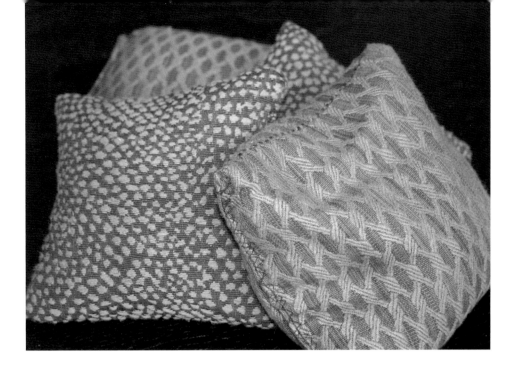

3. environments: changing from indoor to outdoor conditions.

Problem 10b: Designing Access Principles Assignment

This assignment draws attention to certain structural elements that are merely details of a larger structure. Specifically, it deals with the access principles to an enclosure—i.e., doors, drawers, slides, etc. This aspect of an object may seem to be of minor importance and, for this reason, is often dealt with in an incidental manner.

Obviously, all enclosures require an access. The questions are where and what kind. Should it be a surface opening or one of a number of possible split openings (like most refrigerators today)?

And what different positions should the means of access assume? Providing access to an enclosure involves moving parts of the unit from a closed to an open-position. However, the open-position is often ignored or determined by the hardware used. Thus the open-position is often designed by default. This is acceptable in cases where the open-position is only briefly maintained, as in an automobile or refrigerator. In many homes, however, there are doors, for instance, which are never or rarely closed. This fact is usually ignored. In many places where this is the case, the door fits properly in its place, in its closed-position, but in its open-position, it hangs awkwardly in space; this is obviously a default position. The same is true of many enclosures; the hardware (hinges, etc.) determines the open-position.

How should this aspect be dealt with? The door—e.g., of a cabinet—does not have to swing in space just because standard hinges are being considered. With specially designed hardware, a door can be moved from a closed-position to a number of different open positions: below, to the side, above, or even inside the cabinet. This assignment draws attention to this aspect by

fig. 39 (below)
Access Principle,
P. M. Darnell, 1961
fig. 40 (right)
A Director's Chair
Could be Considered a
Type of Compression Chair
fig. 41 (opp. page)
Human Dimensions,
J. D. Robinson, 1961

inducing the student to develop various access principles. It may not always be feasible, from a commercial point of view, to introduce new kinds of accesses, but as in other situations, it is better to deliberately discard an idea than to remain unaware of the possibilities (fig. 39).

Problem 10c: Designing Special Structures

In designing the structure of a chair, it becomes obvious that the floor on which the chair rests is an integral part of the structure. This fact is generally ignored or is simply taken for granted. If it is considered, along with the fact that the weight of a person sitting in a chair tends to pull the chair apart, a structure can be designed which takes both aspects into account.

The compression chair is designed in such a way that the weight of the person sitting in it actually pulls the structural members together via cables below the seat (fig. 40). The joints are

ACCESS PRINCIPLE

PRINCIPLE: 4-BAR LINKAGE
AS SHOWN, THE SYSTEM IS APPLIED TO A
PORTABLE RECORD PLAYER. IT ALLOWS THE TOP
PORTION OF THE CABINET TO BE OPENED WITHOUT
PULLING THE CABINET AWAY FROM THE WALL. THIS
TYPE LID ALLOWS EASIER ACCESS TO THE
TURNTABLE. OTHER APPLICATIONS MIGHT INCLUDE
HORIZANTAL FREEZER LOCKER, DISH WASHER, &
FOOT LOCKER.

I.D. 301 — NO. 6
FALL '61 — P. M. DARNELL

HUMAN DIMENSIONS

STATURE RANGE

RANGE OF EYE HEIGHTS

RANGE OF SEAT HEIGHTS

EYE HEIGHTS, STATURE HEIGHTS, SEAT HEIGHTS OF ADULT MALES

REACH OF SEATED ADULT MALES

TEN-PERCENTILE	AVERAGE	NINETY-PERCENTILE
HEIGHT- 46"	HEIGHT- 52.5"	HEIGHT- 56.5"
EYE HEIGHT- 44.5"	EYE HEIGHT- 48.5"	EYE HEIGHT- 52.5"
SEAT HEIGHT- 12.5"	SEAT HEIGHT- 16.5"	SEAT HEIGHT- 20.0"

SCALE 1½"=1'-0"

ROBINSON J.D.   I.D. 202   FALL 6

all loose, the rods merely resting in the holes of the tubing and are tightened only by weight in conjunction with the rigidity of the floor. The floor does not have to be completely flat. In itself, such a structure may not have any commercial importance, but it serves to alert the student to aspects of a design procedure that are often not at all obvious.

An interesting result of the broad approach followed in this assignment was a chair with a built-in gyroscope, which required only two legs and allowed for rather simple changes in the leaning angle of the chair, forward or backward.

Problem 11: Anthropometry Assignment

This assignment, which serves as a partial introduction to anthropometry, calls for the making of scale templates of the human figure with moveable joints, suitable for making drawings. In order to have templates of different positions and scales available for the class as a whole, each student makes a different template as to the scale and position of the figure (fig. 41).

Since sufficient publications covering the subject of anthropometry are available today, it will not be considered here further.

Problem 12: Unit X Approach Assignment

This assignment is an introduction to an analytic design procedure. The approach represents an effort to establish a more systematic design approach in applying the knowledge dealt with in various preceding assignments. This procedure will be dealt with here only to illustrate its basic nature and its place in the overall educational program. The assignment is system oriented, and calls for the development of a defensive-type object belonging to the transformer group. The basic characteristics of this type of object have been dealt with in Chapter 2.

Some questions may arise about the complexity of this approach, particularly in designing a rather simple object. The time and effort the Unit X approach requires may appear excessive. This, however, is first and foremost an educational exercise, which seeks to draw attention to every detail and aspect of an object and design procedure. It does not suggest that in future every step of this approach is to be followed for every design problem involving a defensive-type object. However, a familiarity with this approach can be useful even in the development of a simple object in that it draws attention to pertinent aspects which otherwise may be neglected. An analogy may serve to illustrate this. When one learns a foreign language, one deals rather extensively and in detail with its different aspects. But when using the language later, one does so without referring constantly to its specific structure and rules. If, however, a question should arise, one can simply and profitably refer to the specifics of the language. This applies equally to the Unit X approach.

In the following, the to-be-designed object will be referred to as the "unit." The results obtained in using such unit are the desired changes or the end-result.

The Phases of the Unit X Approach

The approach consists of eleven successive phases. Phases one through seven are analytical. The subsequent phases synthesize the data obtained in the earlier phases.

Phase 1: Delineation of the problem

Phase 2: Survey of the competitive field

Phase 3: Survey of potential consumers

Phase 4: Establishing the operation cycle

Phase 5: Establishing the inner constants

Phase 6: Establishing the outer constants

Phase 7: Developing operational principles

Phase 8: Developing mechanical principles

Phase 9: Developing technical principles

Phase 10: Developing relational principles

Phase 11: Selecting the final unit

Phase 1: Problem Delineation

It is rather common in stating a design problem to refer to a specific product by name. If taken literally, such an assignment actually only calls for the modification of an existing product. Although at times a given assignment may merely call for modifying an existing product, the procedure presented here refers to the development of a new product.

The defensive type of product is being used, as stated before, to obtain or to control certain changes in other physical entities. Thus, the emphasis here is not on a product but certain desired changes as specified in the delineation of the problem. The task of the designer is to search for and develop the means that can be used to obtain these changes. The changes to be brought about are referred to as the end-result of a utilization process. The assignment would state: "Develop the means that can be used to obtain the following changes," followed by a list of the desired changes and their specifications.

For example, an actual assignment would state: "Develop the means that can be used to facilitate various kinds of writing commonly done in the home. This includes all activities directly related to writing (but it excludes professional writing)." For identifying the project, the assignment is referred to as "operation writing."

Phase 2: The Competitive Field

The competitive field seeks to obtain information regarding the means (products) presently available and used in obtaining the same or similar results outlined in the delineation of the problem, which in one way or another could compete with the product being developed. The study of the competitive field concerns (a) commercially available products; (b) custom-made products; (c) do-it-yourself products; and (d) makeshift arrangements. Analysis of the available means could include the following items:

- availability

- feature analysis

- input-output

- quality of results obtainable

- positive and negative by-results

- spatial requirements

- operation of the means

- operational steps required

- skill and knowledge requirements

- maintenance, by owner, expert, etc. or parts replacement

Phase 3: The Consumer Survey

The consumer survey consists essentially of a questionnaire, which, if required, can be conducted by a specialist in the field.

Phase 4: The Operation Cycle

The operation cycle is an important part of the Unit X approach. It consists of a step-by-step description of the operational procedure followed in obtaining each of the end-results (desired changes) given in the delineation of the problem. The list may begin at any point (possibly at the end of a dormant phase) and continues through all the successive steps, after which the cycle repeats itself. It is a time cycle in that each cycle may be repeated a number of times each day or over any other time span—weekly, monthly, seasonal, etc. The extent of detail in such a cycle depends on the nature and importance of the assignment. This results in a list of the inner constant.

The purpose of this phase is to obtain a list of the various items (inner constants) essential in carrying out each operational step. To illustrate, let us assume that one of the end-results is a signed, mail-ready check. This operation, for instance, requires a checkbook, pen, envelope, stamp, and a supporting surface, etc. These are the inner constants of the unit being developed and are specifically dealt with in the next phase.

Phase 5: The Inner Constants

This phase specifies the physical characteristics of these inner constants: dimension, form, etc. It also establishes the order and the frequency these inner constants are being used.

Phase 6: The Outer Constants

The outer constants are the various aspects external to the unit that have a physical and/or visual relationship to the unit. This refers to all of the potential placements of the unit, including not only the place or places it will be used but also where it will be stored, maintained, transported, etc. The task is to list and specify these various outer constants.

While the type, nature, and the number of the inner constants varies according to the kind of unit being developed, basically, there are three kinds of outer constants—persons, objects and environmental aspects, and prevailing conditions. These may or may not have some bearing on a product; nevertheless, they will always be present (see Chapter 2).

Potential Outer Constants:

• Persons: list and specify person(s) operating the unit; non-operating person(s) having some contact with the unit (maintenance, transportation, etc.); and non-operating person(s) experiencing the unit only (in passing) visually.

• Existing objects and environmental aspects: list and specify all objects and environmental aspects that have a direct physical and/or visual relationship with the unit in its use, stored, maintenance, transit, and disposal positions.

fig. 42 (opp. page)
Automobile Vacuum Cleaner,
W. E. Kinslow, 1959

• Prevailing conditions: list and specify the various prevailing atmospheric, social, and political conditions in which the unit will exist while being operated, stored, transported, and disposed of.

Phase 7: Developing Operational Principles

In this phase and the following, the data obtained in the preceding phases is synthesized. Having established the procedural steps (operation cycle) essential to obtain specified end-results, the task is now to develop the physical means that make the procedure possible.

Having established the characteristics of the inner constants in an earlier phase, we will now bring the inner and outer constants together by arranging the inner constants in accordance with their actual operation (i.e., usage). If a number of different outer constants are being considered (i.e., the specific placement of the unit), this may call for a different arrangement for each these different outer constants.

1. Arranging the inner constants, within the space confinement of the selected outer constants, according to:

a. the different desired end-results

b. the physical characteristics of the inner constants

c. the order they are being used

d. the frequency they are being used

2. Designing by developing the following:

a. the physical means which retains the obtained arrangements

b. the enclosures of the arranged inner constants
within the confines of the selected outer constants

c. the access to the enclosure

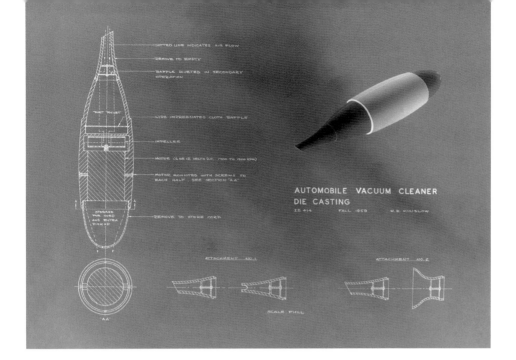

AUTOMOBILE VACUUM CLEANER
DIE CASTING

Phase 8: Developing Mechanical Principles

Develop the mechanics (of the different operational principles)
essential to retain and to provide access to the inner constants.

Phase 9: Developing the Technical Principles

Translate the results of the preceding phases
into specific materials and manufacturing processes.

Phase 10: Developing Relational Principles

Having obtained a number of viable solutions, review and
refine the overall operational and visual aspects of the obtained solutions.

Phase 11: Selecting the Final Unit

The task is now to select from variety of feasible solutions one or perhaps several for commercial consideration. The procedures of the Unit X approach result in a variety of solutions,
depending on the different operational procedures, arrangements of the inner constants,
and the place of operating the unit. This approach does not apply to offensive-type objects,
which also belong to the transformer group.

Problem 13: Products with Engineered Components Assignment

This assignment calls for the designing of a product that includes engineered components—
for instance, a telephone, radio, or vacuum cleaner (fig. 42). The development of the engineered components is, of course, the task of the engineer. The designer generally concentrates

fig. 43 (opp. page left)
Aquarium Showing the Bilateral Approach

on the enclosure of the component, which is sometimes referred to as a "package deal." Such a limited approach may be acceptable in the redesigning of existing products, but in the development of a new product with new engineered components, it is beneficial to involve the designer in the development of the component itself—specifically, in the arrangement of the details of the component and the placement of the controls. This assignment assumes such a situation in which the task of the designer goes beyond that of simply designing the enclosure. When developing engineered components, engineers seem to have a tendency to arrange details in such a way that maximum compactness will be obtained—for instance, by placing the different parts as close together as possible. Maximum compactness primarily aids the manufacturing, storing, and shipping of a unit; utilization is often merely a secondary factor in the arrangement of the details of the components. This aspect has been dealt with in some detail in Chapter 5.

As the liaison between the consumer and industry, the designer's task is not only to package the product but also to relate it appropriately to the user and the environment in which it will be placed. From the point of view of utilization, the arrangement made by the engineer may not be appropriate for the product. The designer may suggest an arrangement of the details that does not ignore compactness but at the same time considers the unit's relationship to environmental aspects.

The student begins by selecting an appropriate product and considering possible rearrangements of the details of the component—obviously, without interfering with their proper functioning. In practice, such a rearrangement would not be arbitrary (although occasionally it is) but would be based on a study of the outer constants—i.e., the different aspects of the environment in which the unit will be used. However, since this is an educational exercise and the intention is to make the student aware of the various possibilities for rearrangement, the student also may choose to devise some arbitrary arrangements, such as obtaining either a thin and elongated form or a tall one.

Three arrangements of the same component are presented in a mock-up form. These represent the so-called volume-form of a component—i.e., its overall form, ignoring minor details. The next step of this procedure is to select one of the three volume-forms for further consideration.

Keeping this volume-form constant, three different enclosing forms are developed, each based on one of the following approaches: (a) the inside-out approach; (b) the outside-in approach; and (c) the bilateral approach.

These approaches have been dealt with in a preceding chapter and only will be briefly discussed here. Form based on the inside-out approach closely follows the general contour of the component (volume-form). The engineered component determines the outer form to a large extent. In contrast, the outside-in approach tends to ignore the volume-form of the component, except in its overall dimensions, actually superimposing a form which bears little or no relationship to the volume-form of the component. The bilateral approach combines both of these aspects, relating the enclosing form appropriately to the component as well as to the environment in which it will be used (fig. 43).

Problem 14: Ecological Approach Assignment

This assignment is a study of the ecological aspects of the man-made object, specifically, the operational interdependency of objects. As has been stated, an object becomes operational (useful) only in conjunction with other objects and environmental aspects, i.e., as an integral part of a system. An assignment may call for the development of a specific object; nevertheless, the designer has to consider the various objects and environmental aspects which make up the system, commonly referred to as the "parameter," or as in this writing the "outer constants." The study is twofold: a) its utilization and b) its manufacturing phase. This includes the complete cycle (its total existence) of the object being developed, its inception, manufacturing, distribution, utilization, discarding, and possible recycling, as well as any changes which may occur within any of these areas.

Problem 15: The Temporal Aspect of the Man-made Object Assignment

This assignment is a study of the successive generation of a given object in time. The student selects an object or area to be investigated.

Like the preceding assignment, this is a further study of the historical development of a certain kind of man-made object. Since each object has its predecessor and successor the questions are: What is the specific nature of the changes which occurred in the past? And what changes can be expected in the near future? The study results in patterning of such development, providing information for the further development of the product considered.

The Application Design Course Series

The Application Design Course Series represents a further step in the teaching of materials and manufacturing processes as part of an educational program for industrial designers.[39] This course series goes beyond mere training in technical proficiency in that it relates such training directly to the designing of products and is thus more precisely tailored to the needs of the industrial designer.

The development and nature of this course series has already been dealt with in Chapter 5. To recall briefly: It began in the pre-Bauhaus period when various shops were introduced in the arts and crafts schools of Germany at the beginning of this century. These shops, however, were not directly related to the curriculum for designers. It was the Bauhaus which made technical training an integral part of design education. Since the aim of the Bauhaus program was to train artist-craftsmen, students were essentially restricted to the one material that was the focus of their major—for instance, wood, metal, or weaving. From today's point of view, this was a rather restricted program for an industrial designer.

Two aspects of the practice of industrial design draw attention to the application approach. First, designers are sometimes faced with the task of finding an appropriate use for a new material, part, or manufacturing process. Second, although we no longer select forms found in nature for commercial consideration, certain kinds of applications remain part of today's products. Not only materials but certain parts and details of most products are not designed specifically for the product in which they are used. These are merely selected. For the designer, it is important to be aware to this fact because these selected aspects often only approximate the actual requirements. Such awareness will give the designer the choice either to accept an application or to design a new one specifically for the product in question.

Of course, the ever-greater variety of materials, parts, components, etc. now readily available to the designer may make specific designs less urgent. Nevertheless, the trend in the overall development of products is towards adjusting products ever more closely to the actual requirements demanded by utilization. This tendency can be observed to a limited degree in the plastics and metal-alloy fields, where certain material compositions are specially designed. In other specialized fields as well—for instance, military and space exploration—where appropriateness is of vital importance, materials and parts are often designed for specific purposes.

Whatever the trend, application—i.e., the selection of certain details in our products—will remain part of most design procedures, despite the fact that it often only approximates the actual requirements. To make designers aware of this, they should be given the choice either to accept a specific application or to design it especially for the product in question.

Some of the common types of applications a designer may have to cope with are as follows:

1. Composition of materials

2. Preforms

a. Sheet materials: plywood, composition board, metal, plastic, etc.

b. Elongated materials: tubing, profiles, extrusion, moldings, etc.

c. Modular materials: bricks, rocks, etc.

3. Standard parts: fasteners, hardware, etc.

4. Components: mechanical, electrical, chemical components,
e.g., electric bulbs, switches, motors, etc.

5. Manufacturing processes: forms resulting from such manufacturing processes
as the potter's wheel, metal turning, the spinning lathe, etc.

6. Utilization

a. abstract forms: geometric, etc.

b. sham forms: a clock in the form of a teapot, etc.

c. natural forms: a table leg in the form of a lion claw, etc.

d. symbolic forms: a product in the form of a symbol

Assignments in the Application Course Series

The course includes a series of lectures covering all major material-transforming processes, such as: wood shaping, wood routing, ceramic, wood laminating, plastic forming, metal casting, blow molding, fiberglass forming, metal and plastic extrusion, tubing forming, sheet-metal forming, roll forming, wood molding, press forming, glass blowing, glass pressing, forging, die casting, injection molding, and impact extrusion.

The aim of this course series is twofold: to introduce the student to a design approach that seeks to make use of an existing means (an often neglected aspect in the education of designers); and to familiarize the student with the major materials and manufacturing processes, emphasizing the creative point of view. This series proceeds from a given means towards an appropriate situation, the reverse of the Situation Design Course Series (figs. 44-46). Each assignment calls

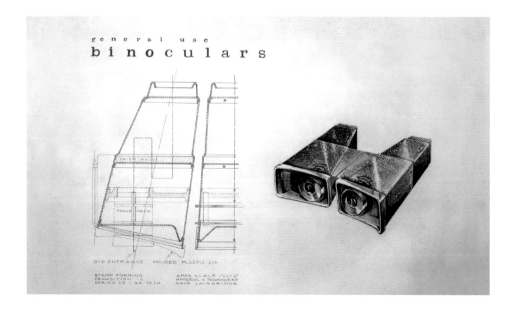

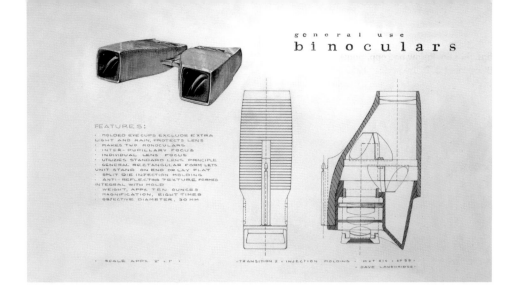

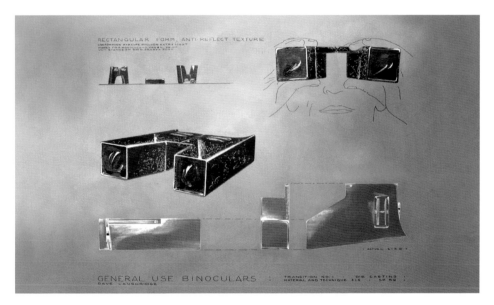

for the designing of a product, using a specified material and a certain manufacturing process. Some of the work is presented in the form of models, some only in renderings. By using models to manipulate various materials and processes, students also gain skills in model-making. Lectures and, whenever possible, demonstrations introduce students to different materials and processes.

In order to familiarize students with the shop facilities, a preliminary assignment calls for the designing and making of simple products in wood, metal, and plastics. This is followed by a series of assignments, each specifying a particular material and manufacturing process. Only a sampling of assignments is dealt with here to indicate the approach. Following the introductory assignments in wood, metal, and plastics, the actual course series begins. The important point in all the following assignments is to think in broad terms about numerous possibilities, avoiding the one-solution approach and going beyond what is already known. A specific kind of product

fig. 47 (below)
Wooden Bowl,
Wood-Turning Assignment
fig. 48 (opp. page)
Wire Door Handle,
Wire Fabrication Project

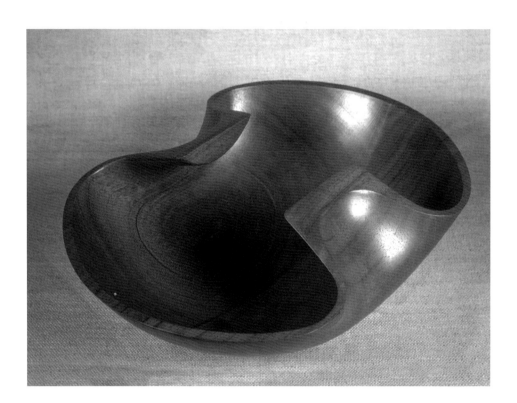

is generally not specified. It is up to the student to apply the specified material in an appropriate manner. The first assignment deals with the joints in our products.

The Joints Assignment

Most of our products consist of a number of parts which are joined together. Although a joint is merely one of many details of a product, it often adds to the product's cost and may influence its outer form. Joints may be referred to as "necessary evils," and there is a tendency to minimize their number or avoid them altogether. Relatively large products are today often made in one piece. Nevertheless, most of our products will continue to have joints. In view of this, it is

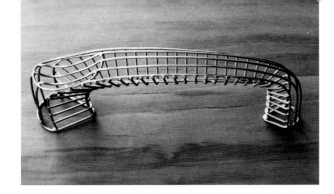

somewhat surprising that design education pays so little attention to them. Standard joints are often used in a routine manner. This may be acceptable in some cases, but when joining affects the outer form of a product, it becomes important to the designer.

This assignment introduces students to the nature and variety of joints. First, the students make some standard joints. Then they design a new joint. It does not have to be a joint that has a practical application. The aim of the assignment is to draw attention to the fact that joints may have some bearing on the form and perhaps also the operation of a product.

The Wood-Turning Assignment

Wood turning may be of minor importance as a commercial enterprise, but like other assignments in this series, it can be useful to further the development of the student's creative faculty.

A demonstration of how to work on the lathe introduces the student to the known possibilities and limitations of wood turning. The instructor demonstrates a face-plate operation by making a simple bowl. The student does not repeat the operation by merely making another bowl. Since the aim is not to train a craftsman but someone able to exploit a given process (in this case, a face-plate operation) in new and different ways, the student is expected to go beyond what has been demonstrated. The aim is to create entirely new forms—forms which the student may not have thought were possible or that he or she could produce. The student is guided in this task by the instructor in accordance with a set procedure (fig. 47).

The Wire Fabrication Assignment

After the student has been introduced to various kinds of wire and their manufacturing processes, he or she is given an assignment to design a product using a specified material, preferably a product presently not made out of wire.

In the Application Course Series, it is generally up to the student to select the kind of product that will be designed. However, since students often have difficulty deciding what kind of product to design, and the time allotment is limited, a particular product is suggested to them, as in this case (fig. 48).

Material Translation Assignment

This is the last assignment in the Application Course Series. It calls for the translation of a given product into another material and/or manufacturing process. At this stage, the student is familiar with all the major materials and manufacturing processes.

There may be various reasons for redesigning a product in a different material and/or with a different manufacturing process. A common one is the desire to improve its performance by taking advantage of a new material or process.

In this assignment, the student selects a certain product and translates it in two essentially different materials and manufacturing processes; e.g., a stamped metal product may be translated into injection molding and perhaps die casting or impact extrusion. The performance of the product is kept essentially the same or may be improved. For the designer, the problem is to exploit the new material and manufacturing process to the best advantage of the product in question.

notes

notes to introduction

1   It must be kept in mind that Professor Bredendieck wrote this over twenty years ago and today all of our permanent Industrial Design faculty conduct sponsored research.—ed.

2   Herbert Bayer, Walter Gropius, Ise Gropius, eds., *Bauhaus Weimar 1919-1925, Dessau 1925-1928* (Stuttgart, Germany: Gerd Hatj, 1955), p. 28; Hereafter referred to as Bayer (1955). Exact reference not confirmed.—ed.

notes to chapter 1

3   For examples of the student work from this period see Walter Gropius et al, *Staatliches Bauhaus Weimar, 1919-1923* (München: Bauhausverlag Weimar, 1923), pp.108-128.

4   Ibid, 127.

5   Herbert Bayer, Walter Gropius, Ise Gropius, eds., *Bauhaus, 1919-1928* (New York: The Museum of Modern Art, 1938), pp. 135-139, 129; First paperbound, 1975, reprinted 1979, 1984, 1986, 1990, Second printing in 1952 by Charles T. Branford Company, Boston; Hereafter referred to as Bayer.

6   Bredendieck's personal recollections.—ed.

7   Bredendieck intended for there to be images of this assignment with glass plates and glass beads; however, the exact images could not be located in his image collection. Throughout the manuscript when images could not be found, the figure was either omitted or another appropriate image was substituted.—ed.

8   László Moholy-Nagy, *The New Vision: Fundamentals of Design, Painting, Sculpture, Architecture*, The New Bauhaus Books Series 1: Gropius and Moholy-Nagy, series editors (New York: W.W. Norton & Company, 1938); First German edition 1928, many subsequent English 1946, 1947, republished 2005 by Dover, Bredendieck referred to pages 48-50 but did not reference an edition, either German or English; Bayer, 123-124.

9   Ibid, 115-118.

10   Moholy-Nagy, *The New Vision*, 56, 57, 92, 93.

11   Bayer, 32.

12   Bredendieck's parenthetical reference read: "Die Form, no. 6, 1930" which referred to the German language magazine *Die Form*.—ed.

13   Moholy-Nagy, *The New Vision*, 21.

14   Moholy-Nagy, I.D. Folders, 1947. Reference as Bredendieck wrote it, unable to locate exact source for the references containing "I.D. Folders."—ed.

15   Moholy-Nagy, *The New Vision*.

16   Moholy-Nagy, I.D. Folder, 1947.

17   Bayer, 26.

18   Moholy-Nagy, I.D. Folder, 1942.

19   Bayer, II6.

20   Moholy-Nagy, I.D. Folder, 1942.

21   Taken from a statement of about 1948, issued by the Visual Design Workshop
      at the Chicago Institute of Design. Reference as Bredendieck wrote it.—ed.

22   Bayer, 28.

23   Ibid, 155-156.

24   William C. Renwick, ASID president, "ASID Newsletter," April-May 1963.

25   Richard S. Latham, ASID, "ASID Newsletter," Jan.-Feb. 1962.

26   Raymond Spilman, FASID, "ASID Newsletter," April 19, 1963.

27   Peter Muller-Munk, FASID, past president of ICSID, Second ICSID
      general assembly in September 1961.

28   Leonardo da Vinci, *A Treatise on Painting*, trans. Jean Francis Rigaud,
      Great Minds Series (New York: Prometheus Books, 2002): 132; Actual quote
      in this translation reads: "Those who become enamoured of the practice of
      the art, without having previously applied to the diligent study of the scientific
      part of it, may be compared to mariners, who put to sea in a ship without
      rudder or compass, and therefore cannot be certain of arriving at the wished-
      for port. Practice must always be founded on good theory; to this, Perspective
      is the guide and entrance, without which nothing can be well done."
      Reference added; it is not known from which source Bredendieck quoted.–Ed.

29   There was no reference for this quote; however, according to art historian
      Gerald Nordland, Josef Albers (1888-1976) authored this in 1949;
      and it appeared in *Ten Variants*, New Haven: Ives-Sillman, Inc., 1967;
      http://www.a-r-t.com/albers.htm#st. (accessed 8 August 2009).—ed.

notes to chapter 3

30   Bredendieck is referring to the *"Manifesto of Tactilism"* (1921)
      by Filippo Tommaso Marinetti.—ed.

notes to chapter 5

31   Bayer (1955), 28, 24.

32   *Idee und Aufbau des Staatlichen Bauhauses Weimar, 1923*. This paragraph
      has not been edited; Bredendieck is referring to Walter Gropius' 1923
      "Idea and Design of the Weimar Bauhaus," which was the manifesto of
      the Bauhaus program of 1919. It is here that Gropius advocates a close
      unity of art and industry.—ed.

33   R.H. McKim, "Aesthetics in Engineering Products," *IDEA Journal* (1963).

34   "The Press: Corbu at Harvard," *Time*, November 30, 1959; Le Corbusier
      did say this in 1959; however, he was quoting Frank Lloyd Wright.—ed.

notes to chapter 6

35   László Moholy-Nagy, "New Approach to Fundamentals of Design," *More Business* (November 1938).

36   Moholy-Nagy, "Briefly, the Objectives," *Bulletin of the Institute of Design,* no date provided by Bredendieck.—ed.

37   See *Life* magazine (16 May 1955).

38   Alex F. Osborn, *Applied Imagination: Principles and Procedures of Creative Problem Solving,* (New York, New York: Charles Scribner's Sons, 1953).

notes to chapter 7

39   One version of the manuscript had this section starting a new chapter; however, following the most complete version and the one provided by his children, I kept The Application Course Series within Chapter 7.—ed.